DISCARD

COLLECTION MANAGEMENT

The Basics of Biology

The Basics of Biology

Carol Leth Stone

Basics of the Hard Sciences
Robert E. Krebs, Series Editor

GREENWOOD PRESS
Westport, Connecticut • London

Library of Congress Cataloging-in-Publication Data

Stone, Carol Leth.
    The basics of biology / Carol Leth Stone.
        p. cm. — (Basics of the hard sciences)
    Includes bibliographical references and index.
    ISBN 0–313–31786–0 (alk. paper)
    1. Biology. 2. Human biology. I. Title. II. Series.
QH307.2.S75   2004
570—dc22   2004008510   _570_

British Library Cataloguing in Publication Data is available.

Library of Congress Catalog Card Number: 2004008510
ISBN: 0–313–31786–0

First published in 2004

Greenwood Press, 88 Post Road West, Westport, CT 06881
An imprint of Greenwood Publishing Group, Inc.
www.greenwood.com

Printed in the United States of America

The paper used in this book complies with the
Permanent Paper Standard issued by the National
Information Standards Organization (Z39.48–1984).

10 9 8 7 6 5 4 3 2 1

In memory of Paul DeHart Hurd, teacher and friend.

Contents

Preface

This book will describe what the living things on our world are like today, how they became what they are, and how their characteristics are passed from one generation to the next. It will also show how living things interact with each other and with their nonliving surroundings.

Never has the need for biology education been greater. We are continually bombarded with new information about environmental problems, bacteriology, genetic engineering, drugs, health, and other biology-related topics. Yet few of us retain enough knowledge about basic biology to make informed decisions or even to read newspaper science articles intelligently.

It is no longer possible—if it ever was—to provide a comprehensive survey of biology, but the material that follows will enable readers to read and evaluate articles and books written for the general public. In addition, the investigations, suggestions for further reading, and list of Web sites can lead readers beyond this basic introduction.

Readers will not become instant experts on biology, of course. This book is designed mainly as a reference work, to be used as needed for understanding the biological articles, books, and news programs that surround us. Readers should achieve at least a level of understanding that will enable them to read biology articles in the *New York Times* or *Scientific American* without becoming confused.

Chapter 1 gives an overview of ecology, providing a general picture of the biosphere and its inhabitants. In Chapter 2, evolution is presented as the process that has resulted in the biosphere of today. Chapter 3 discusses genetics, the molecular basis of variations that are the substrate for evolution. Body systems, especially human systems, are the subject of Chapter 4. In Chapter 5, the major classes of living organisms are surveyed. Completing the cycle, Chapter 6 returns to ecology, with emphasis on human ecology. The investigations or experiments in Chapter 7 have several purposes. Some provide additional concrete examples of ideas in the text; some help readers plan and carry out biological experiments; and some link the information in this book to events in the environment. Rather than being "cookbook" types of investigations, they are open-ended, encouraging readers to ask their own questions, set up experiments, and explore new worlds of information. When definite answers to questions are called for, they are provided (on pages 211–213).

Scientists are introduced throughout the text as part of the history of each field, rather than being relegated to a separate section. Dates of birth and death are given at the first mention of scientists no longer living. The Landmarks in Biology section (pp. 221–226) also provides a timeline showing when individuals made their important contributions to biology.

The appendixes have resources for further reference. They include (1) Investigating Biology on the Internet, a list of the best Web sites for learning more about biology, keyed to the chapter in this book to which they most refer; (2) the Units of Measurements section, which provides conversions for the metric system, the universally used system in science and outside of most of the English-speaking world; and (3) Landmarks in Biology, a timeline of important events in the history of the study of biology.

To maximize the book's usefulness as a reference work, a large glossary and index are provided. Any boldface terms in the text appear in the glossary, as do many other terms. Bold words are usually those whose definitions would slow down some readers but may be needed by others. Finally, a selected bibliography lists some of the printed resources used in researching this book and good books and articles that provide even more suggestions for further study of the topics in biology discussed herein.

Hundreds of sources were used for collecting and verifying the information given here. In addition to biology books and articles, the World Wide Web has been a very useful source of information.

Concepts are presented as the first two parts of the "learning cycle" often used in science education: An example is given first, followed by the general concept. The third part of the cycle (application of the concept to a new example) is up to the reader. It is hoped that readers will recognize the principles given here as they encounter new examples while carrying out investigations, visiting museums, or watching television. More than recognition should result: Readers should also be better able to analyze information about biology, synthesize information from different sources, and evaluate arguments involving biology.

Though this is a reference book and not a textbook, a creative teacher may wish to use it as the basis of a biology curriculum. The outline was carefully developed in accordance with the National Science Education Standards for high school biology. Obviously, this book is shorter than most modern biology textbooks; if it is used in classrooms, students may carry out independent projects to supplement the basic ideas given here.

Acknowledgments

For encouragement, advice, and information of various kinds, I am greatly indebted to the following: Master Gardener Alora Davis; Professor Paul R. Ehrlich, Stanford University; the late Professor Ned K. Johnson, University of California, Berkeley; Richard M. Jones, Head of Library Reader Services, British Medical Association Library; the late Emeritus Professor John A. Moore, University of California, Riverside; Dr. James. N. Shoolery, NMR Applications Chemist; my husband, the late Harold J. Stone; Professor David B. Wake, University of California, Berkeley; and Professor Tim White, University of California, Berkeley. Diane Worden was the indexer. Rae Déjur prepared the illustrations, and development editor Anne Thompson and series editor Robert Krebs waited patiently while I finished the manuscript. Laura Poole was the copyeditor. They all gave me invaluable editorial advice, but all responsibility for errors is mine alone.

The Biosphere

The nonliving parts of Earth can be considered as three realms: the *hydrosphere* (water), the *atmosphere* (air), and the *lithosphere* (rock). The living part is the biosphere.

In 1940 biologist E. F. Ricketts (1897–1948) and novelist John Steinbeck (1902–1968) set out from Monterey, California, in a salmon-fishing boat to collect specimens of animals from the Gulf of California in Mexico. They returned with examples of more than 550 **species,** which Ricketts later used in compiling a vast catalog of animals in that region. (A species is a group of organisms that can form fertile offspring only with each other.) More than that, they returned with a shared philosophical outlook about how all species on Earth, including humans, are linked with one another and with the nonliving environment. Later Ricketts and Steinbeck wrote of their expedition,

Our own interest lay in relationships of animal to animal. . . . it seems apparent that species are only commas in a sentence, that each species is at once the point and the base of a pyramid. . . . One merges into another, groups melt into ecological groups until the time when

what we know as life meets and enters what we think of as non-life: barnacle and rock, rock and earth, earth and tree, tree and rain and air . . . all things are one thing and . . . one thing is all things—plankton, a shimmering phosphorescence on the sea and the spinning planets and an expanding universe, all bound together by the elastic string of time. It is advisable to look from the tide pool to the stars and then back to the tide pool again.[1]

As poetic as that passage is, it is also a valuable summary of ecology, the study of interrelationships among living creatures and the environment. The name comes from the German *oecologie,* a word first used in 1866 by biologist Ernst Haeckel (1834–1919).

FROM ORGANISMS TO ECOSYSTEMS

How can anyone begin to comprehend the rich variety of life on Earth? There are many different ways of learning about life, as well as understanding and appreciating it. An artist, a writer, and a biologist use different modes of study and different ways of reporting their understanding of life.

TOOLS AND TECHNIQUES: THE CURATOR'S TOOLKIT

The dioramas and other exhibits in science museums are only a shallow surface layer of the museums' collections; and at research institutions, there may be no public exhibits at all. The heart of any museum is the collections and records themselves, which act as a tangible library.

Collecting and cataloging animal specimens has come a long way since John James Audubon (1785–1851) shot many of the birds he painted for his famous *Birds of America*. The conservation-minded museum biologists of today are as likely to bring back photos as they are to collect dead animals, though they always make the most of opportunities to collect beached whales or other interesting animals that have died.

The Museum of Vertebrate Zoology (MVZ) at the University of California, Berkeley, is one example of a modern university museum with an enormous collection of animal specimens. Cabinets and glassed-in cases hold about 630,000 items—stuffed animals, tissue samples, study skins, and skeletons—collected since 1908, when enthusiastic explorer Annie Alexander donated money to the university for a museum of birds, mammals, amphibians, and reptiles.

Some of the specimens are holotypes for their species; a *holotype* is the original specimen selected by biologists as the species type and from which the original description was made. Holotypes have the standard characteristics against which newly collected animals can be compared and classified.

Almost as important as the specimens is the MVZ's collection of field notebooks filled with biologists' careful notes about climatic and other environmental conditions taken when they were studying animals in the field. Now comparisons are possible between environmental conditions in a certain area today and a hundred years ago because of those notebooks. Though they have been used as the raw material for writing published articles and books, the notebooks themselves are invaluable tools.

Museum biologists may study the size and geographic range of a population by sampling techniques (counting the number of organisms in a few representative areas and extrapolating that number to a large area) and by tracking. Radios are often used for tracking; the animal is captured and fitted with a radio that transmits its location at all times.

DNA analysis is part of a modern museum's gamut of tools. DNA in microscopic bits of 100-year-old study skins can be analyzed for comparison with that in modern animals or with that in other older study materials.

Over the past few years, the MVZ's collections have been organized as electronic files. This has made it possible to search the records and to make analyses more easily.

It has always been important to make drawings or photos of vertebrates and to measure their tails, limbs, and other anatomical parts. Making these observations is still part of what museum biologists do. However, the MVZ's first director, Joseph Grinnell, was an ornithologist who saw birds in the context of their environments, and his outlook has influenced museum studies ever since. Also, in recent years the MVZ has become more multidisciplinary and has emphasized the ecology and evolution of organisms. This has led to adopting some of the tools and methods of other disciplines.

Although all scientists share a reliance on logic and experimental methods, the methods of a biologist differ from those of a chemist or astronomer. Biologists are especially dependent on describing living things, rather than experimenting, as a way of gathering data.

Many biologists study living things on the molecular level, using elaborate instruments. However, a beginning scientist (or anyone who wants to know more about the natural world) can get an overview of life by looking at whole organisms (living things).

One organism, an individual such as a cat, may live alone for much of its life. It interacts with others of its kind only occasionally, such as during times of fighting for territory or of courtship and mating.

Most organisms, though, live in groups. When a group of individuals of the same kind lives in the same area, the group is called a population. The various populations in the area make up a community of organisms that interact in many ways. A community in the Everglades of Florida could include alligators, saw grass, deer, and other plants and animals.

What an organism eats and is eaten by determine part of its niche, or way of life. Each species has a unique niche, shared by no other species. This important concept was developed by American biologist Joseph Grinnell (1877–1939) in 1917.

Where the organism lives is its habitat. For example, a desert rattlesnake's niche may include eating mice and being preyed on by hawks, but its habitat is the desert.

An organism's surroundings, or environment, are both living and nonliving. The living portion (the biotic environment) is the rest of the community of organisms; the nonliving part (the abiotic environment) is made up of rocks, water, and air.

Taking everything together—a community of organisms and their abiotic environment—you have an ecosystem for an area. Ecosystems must have arbitrary limits, because there are many overlaps among them; air and water can travel, as can seeds and animals, crossing ecosystem boundaries. An ecosystem in a pond is quite different from the ecosystem on land next to the pond, but some birds, frogs, and plants may belong to both ecosystems. Usually an ecosystem is defined as the system within a given geographic area with a particular kind of climate (the long-term weather conditions in an area, especially temperature and precipitation).

Biomes

Certain large ecosystems are called **biomes.** These extend over large areas that have generally similar climates. In North America, for example, we have the deciduous forest, taiga or boreal forest, grassland, hot desert, cool desert, chaparral, tropical rain forest, and tundra biomes. Similar biomes are found in equivalent climates on other continents.

Geography's connection to climate is complex, but a general pattern emerges. The movement of air from northwest to southeast across the North American continent has an indirect effect on biomes. As moist air from the Pacific Ocean reaches the West Coast, it brings fog and rain to coastal areas. The now-drier air moves inland until it reaches the mountains, where it rises. As it rises, the air is cooled; eventually it loses moisture as precipitation on the leeward slopes of the mountains. By the time it reaches the Great Plains east of the Rocky Mountains or the desert areas of the Southwest, the air is dry again. It continues moving eastward above the prairies, gradually picking up moisture. By the time it reaches the Great Lakes states, the air is wet enough to produce considerable rainfall. The movements of wind and water along the East Coast have their own effects on temperature and precipitation.

Temperature and precipitation are both important components of climate, but precipitation is generally considered the main determining factor of vegetation type. This idea was originated by Danish botanist Eugene Warming (1841–1924), who wrote *Ecology of Plants* in 1896 and has been called the father of ecology.

Figure 1.1
North and Central America Biomes Map.

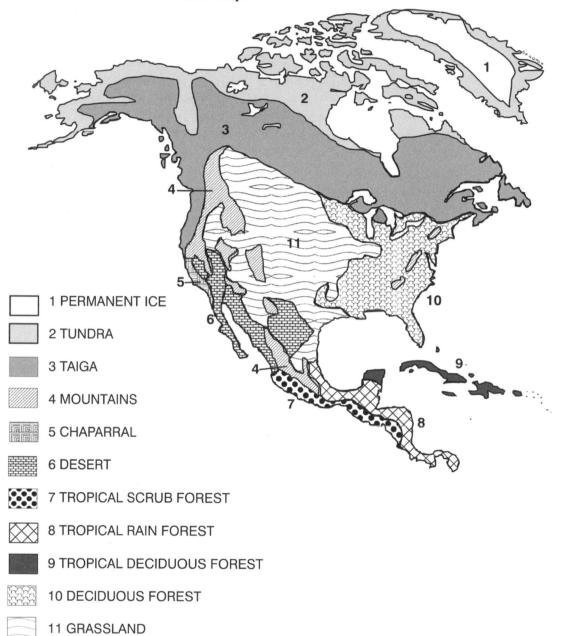

1 PERMANENT ICE

2 TUNDRA

3 TAIGA

4 MOUNTAINS

5 CHAPARRAL

6 DESERT

7 TROPICAL SCRUB FOREST

8 TROPICAL RAIN FOREST

9 TROPICAL DECIDUOUS FOREST

10 DECIDUOUS FOREST

11 GRASSLAND

Middle-Latitude Deciduous Forest

In the lower peninsula of Michigan, spring comes slowly. In March the heavy snows give way to slush and then to running water, which makes growth possible. By May apple and other fruit trees are in blossom, and a warm, moist summer follows. In autumn the leaves of oaks, maples, and other trees take on their fall colors, then drop to the ground as winter approaches. By the end of November, snow is usually falling, and winter is beginning.

This is part of the deciduous forest that spreads across much of the Midwest and East in the United States. Well-defined seasons with distinct winters are seen in this biome. Because most of the trees are broad-leaved and deciduous (they lose their leaves at some time during the year), they look quite different during the four seasons. A moderate climate and a growing season of four to six frost-free months distinguish these forests.

The temperature varies from $-30°C$ to $30°C$ (about $-22°F$ to $86°F$). Precipitation is distributed evenly throughout the year in the form of 75–150 cm (29–58 inches) of rainfall. The soil in these forests is fertile, enriched with decaying **litter.**

Trees of the deciduous forest have broad leaves that are lost annually; they include oak, hickory, hemlock, beech, maple, cottonwood, basswood, elm, and willow. The moderately dense leaf **canopy** allows light to penetrate, allowing bushes and small plants to grow beneath the trees. Spring-flowering herbs are common. The plants in turn are food for a great variety of animals: squirrels, skunks, rabbits, birds, and deer. Lynx, foxes, timber wolves, and black bears are predators where human habitation allows them.

Many of the original deciduous forests were cut for logs and fuel early in U.S. history as settlers of European origin moved west across the continent. Today there are only a few scattered remnants of virgin forest in remote Appalachian regions and larger areas of second-growth forest.

Grasslands

Much of the rural landscape in Iowa and Illinois today consists of flat land covered with corn or soybeans. In a few pockets of native plants near railroad tracks or other "neglected" areas, there are tall grasses.

West of the deciduous forests, trees and large **shrubs** gradually give way to grasses. In Indiana and Illinois, they are the tall plants of the tall-grass prairie; farther west, they gradually give way to the lower plants of the short-grass prairie. Still farther west, prairies become **plains.** Plains and prairies are all included in the grassland biome.

Temperate grasslands have hot summers and cold winters. Precipitation in the temperate grasslands is moderate (the annual average is about 51–89 cm [20–35 inches]), usually occurring in the late spring and early summer. The annual temperature range, in contrast, is great: Summer temperatures can be higher than $38°C$ ($100°F$), but winter temperatures can be as low as $-40°C$ ($-40°F$).

The amount of annual rainfall influences the height of grassland vegetation, with taller grasses in wetter, eastern regions. Seasonal droughts and occasional fires may occur, reducing the plants to their roots. The soil of the temperate grasslands is deep and dark with fertile upper layers. It is nutrient-rich from the growth and decay of deep, branching grass roots that hold the soil together and provide food for living plants. Each species of grass grows best in a particular environment (determined by temperature, rainfall, and soil conditions). Seasonal droughts, occasional fires, and grazing by large mammals all prevent woody shrubs and trees from invading and becoming established. However, a few trees, such as cottonwoods, oaks, and willows, grow in river valleys; some small plants grow among the grasses. There are a few hundred species of flowering plants. The various species of grasses include purple needle grass, blue grama, buffalo grass, and galleta. Pioneers crossing the grasslands in their covered wagons were struck by the beauty of the prairie flowers— asters, blazing stars, coneflowers, goldenrod, sunflowers, clovers, psoraleas, and wild indigos.

The mammals in various parts of the grassland include wild horses, wolves, prairie dogs, jack rabbits, deer, mice, coyotes, bison, foxes, skunks, and badgers. Blackbirds, grouses, meadowlarks, quail, sparrows, hawks, and owls fly above the grasses. Snakes, grasshoppers, leafhoppers, and spiders are other community members.

Few natural prairie regions remain; most have been turned into farms or grazing land because they are flat, treeless, covered with grass, and have rich soil.

Hot Desert

Arizona was sparsely settled for many years because of the intense heat. Since air conditioning and irrigation have come to the area, people have been able to live there comfortably. Arizona typifies the biome called the hot desert, where broad expanses of sand support only the hardiest of plants and animals.

Deserts are found on about one-fifth of the Earth's surface, where rainfall is often much less than 50 cm (20 inches) per year. Most deserts have many plants with special adaptations as well as specialized animals. Soils often have abundant nutrients, needing only water to become very productive. Thus, when winter or summer "monsoon" rains arrive, a desert can suddenly bloom with myriad flowers. Occasion-

ally fires, cold weather, or sudden (infrequent) intense rains may cause flooding.

The hot deserts are generally warm at all times and very hot in the summer. Little rain falls in the winters. Temperatures reach extreme highs and lows daily because the atmosphere contains little humidity to block the sun's rays; desert surfaces receive a little more than twice the solar radiation received by humid regions and lose almost twice as much heat at night. Typical annual temperatures range from 20°C to 25°C (68°F to 77°F), but the extreme maximum may exceed 49°C (120°F). Temperatures sometimes drop as low as –18°C (0°F).

Rainfall is usually very low; it may be concentrated in short periods between long droughts. Evaporation rates regularly exceed rainfall rates, and sometimes rain evaporates before even reaching the ground. Rainfall in American deserts is higher— almost 28 cm (11 inches) a year higher— than in deserts in some parts of the world.

Soils are coarse-textured and shallow with good drainage and no subsurface water. They are coarse because there is little chemical **weathering;** fine dust and sand particles are blown elsewhere, leaving rocks and gravel behind. On hot days, whirlwinds (small, whirling windstorms) may appear.

Plants are mainly low-lying shrubs and short woody trees that are adapted to the heat and dryness. They include yuccas, turpentine bush, ephedras, prickly pears, false mesquite, ocotillo, sotol, agaves, and brittlebush. Most leaves are small, thick, and covered with a thick cuticle (outer layer). In cacti, the leaves are reduced to traces. Some leaves open their stomata (microscopic openings that allow for gas exchange) only at night, when evaporation rates are lowest.

Few large mammals live in deserts because many of them are not capable of storing water and withstanding the heat. They are also too large to hide under objects like rocks or to stay underground, out of the sun. Desert mammals are usually small, such as the kangaroo mice of North American deserts. The dominant animals of hot deserts are reptiles and other nonmammals—insects, arachnids, reptiles, and birds. Many desert animals survive by being nocturnal (active at night) **carnivores.** They stay inactive in protected hideaways during the hot day and come out to forage when the desert is cooler.

Cool Desert

Just east of the Sierra Nevada Mountains, where clouds have lost precipitation as snow or rain, the state of Nevada is quite dry. Though the desert there looks bare, like the hot desert farther south, the climate is different. This is a cool desert.

Cool deserts are characterized by cold winters with much snow and rain falling during the winter and occasionally over the summer. In the United States they are found along the western edge of the Great Plains. Cool deserts have short, moist, moderately warm summers with fairly long, cold winters. The mean temperature in winter is between −2°C (28°F) and 4°C (39°F); in summer, between 21°C (70°F) and 26°C (79°F). The mean annual precipitation ranges from 15 cm to 26 cm (6–10 inches). The heaviest rainfall of the spring is usually in April or May. Rainfall is also heavy in autumn in some areas.

The desert soil is heavy, **silty,** and salty. It contains **alluvial fans,** where most of the salt has been leached out because the soil is relatively porous and has good drainage.

Cool-desert plants are widely scattered with large areas of bare ground. The height of shrubs varies between 15 cm (6 inches) and 122 cm (48 inches). The main plants are deciduous and have spiny leaves.

Cool-desert animals include jack rabbits, kangaroo rats, kangaroo mice, pocket mice, grasshopper mice, antelope squirrels, badgers, kit foxes, and coyotes. Several lizards do some burrowing and moving of soil. Deer are found only in winter.

Boreal Forests

Canada's province of Quebec exemplifies what is sometimes called the "spruce–moose biome," the boreal forest. Large evergreen forests are home to wolves, black bears, moose, lynx, and other mammals. Beavers build their dams in the rivers, where a variety of fish swim.

Boreal forests, also known by their Russian name of taiga, represent the largest terrestrial biome. Boreal forests can be found in a broad belt between the latitudes of 50° and 60° north across Eurasia and North America—two-thirds in Siberia, and the rest in Scandinavia, Alaska, and Canada. Some boreal forests grow in northern Wisconsin, Michigan, and Maine. Summers are short, moist, and moderately warm; winters are long, cold, and dry. The growing season is 130 days long.

Temperatures here are very low, ranging from −20°C to 20°C (about −4°F to 68°F). Precipitation (primarily in the form of snow) ranges from 16 to 40 inches (40–100 cm) annually. Soil in the boreal forest is thin, nutrient-poor, and **acidic.**

The trees are mostly cold-tolerant evergreen conifers with needlelike leaves, such as pine, fir, and spruce. Because the canopy permits only a little light to penetrate, **understory** plants are limited.

Animals include woodpeckers, hawks, moose, bear, weasel, lynx, fox, wolf, deer, hares, chipmunks, shrews, and bats.

Tropical Forests

Visitors to Hawaii, the only U.S. state to have tropical rain forests, soon become accustomed to frequent rains. The sun usually is shining brightly during or soon after the warm rain falls, so few people bother with umbrellas, and spectacular rainbows are common.

Most of Hawaii's plants today have been adopted from other parts of the world, and the few natives that remain grow in remote valleys. At one time, however, about 1,700 unique plant species grew there. Birds include the Hawaiian honeycreepers, a goose, two kinds of hawk, a type of fly-catcher, and two kinds of duck.

The greatest diversity of species of any biome is found in the tropical forests. They occur near the equator, within the area bounded by latitudes 23.5° north and 23.5° south. Some are in Central America and parts of Asia, Hawaii, and South America. Tropical forests have only two distinct seasons: rainy and dry. There is no winter. Day and night are each about 12 hours long throughout the year. Temperature is on average 20–25°C (68–77°F) and varies little: The average temperature during the three warmest months is no more than 5° higher than that during the three coldest months. Precipitation, too, is evenly distributed, with the annual rainfall exceeding 200 cm (79 inches). Further subdivisions of this group are based on seasonal distribution of rainfall: For example, in the evergreen rain forest, there is no dry season.

Tropical forest soil is nutrient-poor and acidic. Plant matter decays rapidly, and soils are heavily leached. Most of the trees are evergreen, with large, dark green leaves. The forest canopy is multilayered and continuous, allowing little light to penetrate.

Plants are highly diverse; a square kilometer may contain as many as 100 different tree species. Trees are 25–35 m (82–115 feet) tall, with buttressed trunks and shallow roots. Smaller plants, such as orchids, bromeliads, vines (lianas), ferns, mosses, and palms, are present.

Animals include numerous and varied birds, bats, small mammals, and insects.

Tundra

If Hawaii is at the highest extreme of heat in the United States, Alaska is at the lowest. Much of the state is tundra, the coldest of all the biomes. The name comes from the Russian *tunturia,* meaning treeless plain. It is noted for extremely low temperatures, little precipitation, poor nutrients, short growing seasons, and landscapes molded by frost. Dead organic material functions as a nutrient pool. The two major nutrients are nitrogen and phosphorus. Nitrogen is added to the soil by **biological fixation,** and phosphorus by precipitation.

Because of the very cold climate, few organisms can survive here, and there is little diversity of plants and animals. Drainage through the soil is also limited. The season of growth and reproduction is short. Populations may rise or fall quickly in number, partly in response to changes in weather.

ARCTIC TUNDRA

Tundras are of two types: arctic and alpine. In the Northern Hemisphere, arctic tundra encircles the North Pole and in some places extends south into the coniferous forests of the taiga. The arctic is known for its cold, desertlike conditions, with the growing season ranging from 50 to 60 days. The average winter temperature is –34°C (–29°F), but the average summer temperature is 3°C (31°F) to 12°C (54°F), allowing this biome to sustain life. Yearly precipitation, which includes melting snow, is 15–25 cm (6–9 inches). Soil is formed slowly. Permafrost, a layer of frozen subsoil consisting mostly of gravel and finer material, is formed at variable depths below the surface. When water has saturated the upper soil, it pools on the surface to form bogs and ponds that provide moisture for plants. As a result, plants of the arctic tundra have no deep root systems. However, a wide variety of plants with shallow roots can resist the cold climate.

The 1,700 kinds of plants in the arctic and subarctic include low shrubs, sedges, reindeer mosses, liverworts, and grasses; 400 kinds of flowering plants; and **lichens.** All of the plants are adapted to high, cold winds, as they are short and grow in groups. They can carry out photosynthesis at low temperatures and low light intensities. The growing seasons are short, and most plants reproduce by budding and division rather than by flowering. They are protected by the snow during the winter.

Mammals include herbivores (lemmings, caribou, arctic hares, voles, and squirrels) and carnivores (wolves, arctic foxes, and polar bears). Migratory species— falcons, ravens, snow buntings, loons, sandpipers, terns, ravens, snow birds, and various species of gulls—represent the birds, and cod, salmon, flatfish, and trout, the fish.

Animals have **adapted** by anatomy or behavior for surviving long, cold winters. Many have additional insulation, such as fat. During the winter many animals **hibernate** or migrate south because food is scarce. During the summer they breed and raise their young quickly. Reptiles and amphibians are few or absent because of the extremely cold temperatures, but insects such as moths, mosquitoes, grasshoppers, flies, blackflies, and arctic bumblebees are found here. The continual **immigration** and **emigration** lead to shifting population sizes.

ALPINE TUNDRA

Alpine tundra is located above the timber line on mountains throughout the world. The growing season is approximately 180 days, with annual temperatures averaging from about −70°C (−94°F) in winter to 12°C (54°F) in summer. The nighttime temperature is usually below freezing throughout the year. Precipitation is from 15 cm to 25 cm (6–10 inches) a year. Unlike that in arctic tundra, the soil in the alpine is well drained. The plants, all of which are very small, include tussock grasses, small-leafed shrubs, dwarf trees, and heaths. Even the willows are only about 8 cm (3 inches) high.

Animals living in the alpine tundra are also well adapted to the frigid conditions. Mammals are such animals as pikas, marmots, mountain goats, sheep, elk; birds are grouselike species; and insects are represented by springtails, beetles, grasshoppers, and butterflies.

Chaparral

Anyone who has watched movie westerns knows what the Old West looked like from the viewpoint of Hollywood. In the area north of Los Angeles, hills are mostly bare but have scattered small trees and bushes. This is chaparral.

Chaparral is a small biome in the western United States; its climate is similar to that around the Mediterranean Sea. The temperature varies little, ranging from about 10°C (50°F) in winter to 24°C (75°F) in summer. Animals and plants here have adapted to long, dry summers and rainy winters. The average annual precipitation is only about 38 cm (15 inches). As adaptations for surviving the dry season, plants are mainly small trees and shrubs that have small evergreen leaves. The coast live oak may be 6–25 m (20–82 feet) high and has a wide canopy, but the leaves are small. Also, many leaves of oaks and other plants are coated with a waxy material. Another adaptation to this climate is the thick underground stem of some plants, which can die back above ground but survive the summer. Similarly, animals may survive the dry season by **estivating** until winter. The chaparral rodents and reptiles have some adaptations like those found in desert animals.

Aquatic Ecosystems

Water covers nearly 75% of the Earth's surface. Aquatic ecosystems are home to numerous species of animals, large and small, and plants. This is also where life started billions of years ago. Today, no life would be able to sustain itself without water. Earth's surface would be rocky and barren.

The aquatic biome is of two basic types, fresh water (ponds, lakes, and rivers) and salt water (oceans and estuaries). The only thing that all the areas have in common is water.

FRESH WATER

Fresh water has a low salt concentration, usually less than 1%. Plants and animals are adjusted to the low salt content and would not be able to survive in areas having high salt concentrations (estuaries or the ocean). There are different types of freshwater regions: ponds and lakes, streams and rivers, and wetlands.

Ponds and lakes range in area from just a few square meters to thousands of square kilometers. Scattered throughout the Earth, many of the first lakes evolved during the Pleistocene Ice Age. Though ponds may be temporary, lasting just a month or two, lakes last for many years.

Three zones, determined by distance from the shoreline and by depth, are found in ponds and lakes. First, shallowest and nearest the shore, is the littoral zone. It is the warmest, being well-lit throughout, and contains diverse species of insects, clams, fishes, amphibians, and algae. Some insects pass through their egg and **larval** stages in the littoral zone. Turtles, ducks, and snakes may live here.

The open water farther from the shore is called the limnetic zone. Because it also receives much light, it has a large amount of plankton (small floating plants and animals). These are the first stages of many aquatic food chains. Many kinds of freshwater fish live here.

The deepest water is the profundal zone, which is also the darkest, coolest, and densest. **Bacteria** and fungi live here, feeding on the dead algae that drift down from the limnetic zone.

Temperature varies according to a pond or lake's size and depth, as well as its latitude and the season. During the spring and fall, the warmer top layers mix with the colder bottom layers because of wind action. This mixture leads to a uniform temperature of about 4°C (39°F) and distributes oxygen throughout the lake.

Flowing from headwaters (springs, melting snow, or lakes) to the mouths of **channels** or oceans are streams and rivers. These change characteristics during their downstream journey, becoming warmer and losing oxygen along the way. They also collect sediments and support more fish that can get along with less oxygen as they near the mouth.

The greatest diversity of plants and animals in moving water is in the middle section between the source and the mouth.

SALT WATER

Oceans are the largest of all ecosystems. They have zones based on depth and distance from the shore, comparable to the zones in lakes and ponds.

In the intertidal zone, which may be a broad sandy beach or a steep rocky shore, live the plants and animals that can survive the harsh and continually shifting conditions of life between ocean and land. This zone is submerged by tides sometimes and exposed at other times. Barnacles, shore-

Figure 1.2
Oceanic Zones.

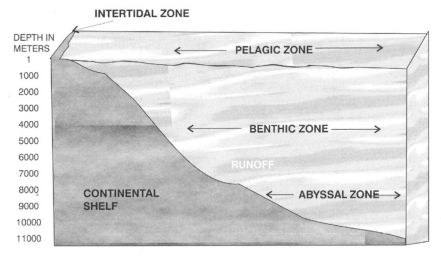

birds, crabs, clams, and starfish live there. Algae and seaweed may be present but not rooted plants.

The open ocean is the pelagic zone. In general it is cold. Seaweed and plankton float on the water's surface, supporting a community that includes fish and mammals, such as whales and dolphins.

Below the pelagic zone is the benthic zone, where the water is colder and darker than in the pelagic zone. Seaweed is the only plant life. Bacteria, fungi, sponges, and fishes are among the other organisms here.

The abyssal zone is in the deepest part of the ocean, where it is very cold and dark. There is a high amount of oxygen but few nutrients. The midocean ridges at the bottom of this zone contain hydrothermal vents, from which hydrogen sulfide and other minerals bubble up into the water. Chemosynthetic bacteria use the energy in these **compounds** to manufacture food, becoming the bases of an oceanic food web, much as photosynthetic organisms support

life elsewhere. Small invertebrates and fishes are supported by the chemosynthetic bacteria.

ESTUARIES AND WETLANDS

Not considered freshwater ecosystems, wetlands are areas covered with standing water where aquatic plants grow. Marshes, swamps, and bogs are all wetlands. They have the greatest diversity of species of all ecosystems. Plants include pond lilies, sedges, cypress, cattails, tamarack, and black spruce. Among the animals are amphibians, reptiles, wading birds, and small mammals.

In places where fresh water joins the salty ocean (estuaries), organisms must be able to tolerate the level of salt found there. Plants and animals such as pickleweed, shrimp, and crabs have special adaptations to salt water. Wading and swimming birds live in estuaries or stop there during migrations. Many worms, oysters, and crabs live in these areas.

Links within the Living World

If you look at a landscape or seascape, it seems to be composed of many separate organisms and abiotic objects. On a molecular level, however, the separate things blend into each other. The building blocks of matter are continually on the move, and they may temporarily join many different parts of the biotic and abiotic environments.

The Chemistry of Organisms

Like all matter in the universe, organisms are made up of chemicals. The simplest chemicals are elements, such as carbon (C) and hydrogen (H). A single atom is the smallest part of an element. Atoms may combine with each other to form molecules; a molecule is the smallest part of a compound. For example, water (H_2O, or HOH) is a compound made of the elements hydrogen and oxygen (O).

The matter in organisms consists mainly of water with other compounds in various combinations suspended in it. Most of these compounds are organic (that is, they contain the element carbon). The elements hydrogen (H), oxygen (O), carbon (C), and nitrogen (N) are the most abundant elements in organisms. Smaller amounts of other elements, such as sulfur (S), iodine (I), and phosphorus (P), are also present.

All living things are composed of the compounds proteins, fats, carbohydrates, and nucleic acids. Proteins, the compounds that make up muscles, consist of long chains of **amino acids.** Every amino acid has the group $-NH_2$ at one end, and the group –COOH at the other. Between those groups are elements that vary according to which amino acid is involved. For instance, the amino acid glycine has the formula NH_2CH_2COOH, and alanine has the formula NH_2CHCH_3COOH.

Fats are the compounds in which animals store food energy. Chains of C and H atoms are linked into long molecules in fats.

Carbohydrates are the sugars and starches made by plants. As their name implies, carbohydrates are made of carbon, hydrogen, and oxygen atoms. When animals eat plants, they take in carbohydrates and break them down for energy by the process of **respiration.**

Chemical compounds may be acidic, basic, or neutral, depending on their electrical charge. Acids have a positive charge because of a positively charged atom (that is, an ion) of hydrogen. Bases have a negative charge because of a negatively charged hydroxide (–OH) group. Neutral compounds have no electrical charge. Whether a compound is charged and whether it is positive or negative determine many of the compound's **properties.**

When water is ionized, the HOH molecules separate into H^+ and OH^- ions:

$$HOH \rightleftharpoons H^+ \text{(hydrogen ions)}$$
$$+ OH^- \text{(hydroxide ions)}$$

(The double arrow shows that the ions also may recombine into HOH molecules.)

For plant growth, the elements C, H, and O (in the forms of CO_2 and H_2O) are necessary for making sugar. In addition, the elements N, P, S, potassium (K), calcium (Ca), magnesium (Mg), iron (Fe), manganese (Mn), zinc (Zn), copper (Cu), molybdenum (Mo), boron (B), and chlorine (Cl) are needed. Plants get them by absorbing the ions from soil water through their roots. Other elements—such as silicon (Si) and aluminum (Al)—are usually present in plants but are not generally considered essential. Silicon, however, has been shown to enhance the growth of many crops. Aluminum, in contrast, is actually toxic to plants.

Compounds and elements move through the living world as plants make food and as other organisms consume it.

Food Chains

In terms of food, organisms fall into three main categories—producers, consumers, and decomposers. Producers are the plants and microscopic organisms that make food from carbon dioxide and water, in a **chemical reaction** that is powered by sunlight or chemical energy. During that reaction, producers make a simple sugar compound (glucose) that may combine with other materials to form more complex substances, such as starches and proteins. These compounds are used by the producer for the matter and energy they need.

Consumers cannot make food themselves; they feed on producers (or on other consumers) and use the matter and energy stored in them. Consumers that feed on producers are called first-order consumers; those that feed on first-order consumers are called second-order con-

sumers; and so on up the line. Any series of organisms beginning with producers and continuing through one or more consumers is a food chain. A food chain always begins with a producer; the number of consumers may vary, though. One simple food chain looks like this:

grass → cow → human
producer → first-order consumer → second-order consumer

No organism is a link in just one food chain. Depending on its niche, each individual affects many other organisms and the abiotic environment. In any region there are cross-connections among food chains, forming a large food web. So removing a link in the web may have effects that reverberate throughout the entire web, not just one food chain. For instance, a food web in the Beaufort Sea in the Arctic Ocean includes polar bears that feed on ringed seals, which are also food for Arctic foxes. The seals eat Arctic cod

---

**APPLICATION TO EVERYDAY LIFE FEATURE:
PROS AND CONS OF BIOLOGICAL CONTROL**

The wild silk moth (*Hyalophora cecropia*), a beautiful moth that can grow to 6 inches across, is disappearing from the northeastern United States. Though habitat loss and pesticides were possible causes of the decline, biologists now think the moth is a victim of predation by a parasitic fly, *Compsilura concinnata* (Yoon 2001).

The fly is not native to the United States but was introduced from Europe in an effort to control the population of the gypsy moth, a pest that has devastated forests in the Northeast and has spread across most of the country. (Introducing parasites as part of

biological control has become a favorite alternative to using dangerous pesticides.) However, the fly not only had a questionable success in controlling gypsy moths, but also is attacking the wild silk moth and other native species. Once introduced into a new area, a parasite usually becomes a permanent part of the community.

Biological control has become increasingly accepted in agriculture and gardening, seemingly to offer the pest-reducing features of pesticides without the dangers of potentially toxic chemicals. Results such as this may lead to reducing these controls.

Source: Yoon, C. K. 2001."When Biological Control Gets Out of Control." *New York Times,* 6 March, p. D3.

and carnivorous zooplankton, small shrimp-like organisms that are also food for bow-head whales and that eat herbivorous zooplankton that feed on diatoms. The cod eat smaller fish and carnivorous zooplankton. If the seals fell victim to a virus and died off, the polar bears and foxes would have less food. Thus freed of predation, the cod could multiply well.

Cycles of Matter and Energy

When organisms die, decomposers, such as bacteria and molds, use the chemical matter and energy they contain for food. As decomposition proceeds, the compounds in the dead organisms are broken down to chemicals that proceed to the non-living environment. Producers can use them again to make food and new living material, which completes a cycle of food production

and use. The cycle is repeated over and over through the same organisms as well as new ones.

In contrast to the recycled matter, the energy used at each step in a food chain is not recycled. Transfer of energy is a one-way street: Producers must always take in new energy to begin a new food chain. Energy is used by the organisms or dissipated as heat. The energy available to each consumer in a food chain is much less than what was available to the organism it fed on; in general, only about a tenth of the incoming energy can be transferred to the next step in a food chain.

Links to the Nonliving World

Just as matter moves from one part of the biotic world to another, it moves between the abiotic and biotic environments. The same atom may be part of an organism at one time and part of a rock or a river at another time, for instance.

Soil

Soil lies at the border between the biosphere and the rock beneath it. The first step in soil formation is the **weathering** of rock into smaller and smaller pieces. Soil may be sandy and beige, clayey and red, heavy and black, depending on where it is found. In any area, soil has a distinctive profile when seen in cross-section. The profile is made up of a series of distinct horizontal layers.

Soil has three chief inorganic ingredients: silt, sand, and clay. The proportions of these determine what the soil looks and feels like, its drainage properties, and what kinds of plants can grow in it. Sand is made of rock particles that are 0.2–2 mm in diameter and are often quartz (silicon dioxide, SiO_2). Silt is like sand chemically, but the particles are finer, 0.02–0.2 mm in diameter. Clay is made of aluminum silicates (such as kaolinite,

Figure 1.3
An Energy Pyramid. The energy transferred at each level of an energy pyramid is considerably less than at the previous level.

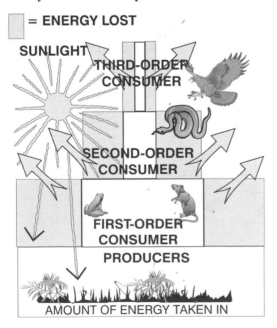

□ = ENERGY LOST

SUNLIGHT

THIRD-ORDER CONSUMER

SECOND-ORDER CONSUMER

FIRST-ORDER CONSUMER

PRODUCERS

AMOUNT OF ENERGY TAKEN IN

$Al_2O_3 \cdot 2SiO_2 \cdot 2H_2O$) that are hydrated (combined with water); particles of clay are very fine, < 0.0005 to < 0.002 mm in diameter.

As weathering proceeds, eventually some primitive plants may grow in the water around the rock particles. Small animals can feed on the plants. As the plants and animals grow and die, their breakdown products combine with the rock particles to become soil. Over years, a long series of reactions such as these may change a bare area into one so fertile it will support a forest.[2]

The minerals in soil water are replenished by being washed in weathered rocks and by **absorption** from the air above the soil. Nitrogen, an especially important mineral for plants, comes in from several sources, including being washed in **nitrous acid** and **nitric acid** by rain or snow. Rain also washes in ammonia (a nitrogen compound) and organic nitrogen from the air.

The chief source of nitrogen for most plants is nitrogen fixation, the conversion of nitrogen to nitrogen compounds by soil organisms. Some nitrogen-fixing organisms live on plant roots, especially those of **legumes;** others live independently.

The organic part of soil is humus. Together with nonliving materials, humus makes up the layer of soil called topsoil. Topsoil is valuable economically for the sustenance of agricultural crops. Wherever plants grow, they drop leaves and seeds and eventually die. All this organic matter decays and becomes humus as the result of breakdown by bacteria and other **microbes.** The products of the decay, complex carbohydrates and proteins, are further broken down into smaller molecules.

Humus varies in different biomes; in a tropical rain forest, most of the humus coming from falling leaves is concentrated in the upper few inches of the soil. In a grassland, the many decaying plant roots lead to humus that is distributed to a greater depth.

In desert areas having little plant life, little humus is found at any depth.

The United States can be divided roughly into western and eastern halves by a line extending from Minnesota through Texas. East of that line, the climate is generally moist; west of it, the climate is drier. In humid biomes, such as the deciduous forest, water moves through the soil quickly, carrying calcium and magnesium ions with it. These ions are replaced by hydrogen ions, lowering the soil's pH. Thus, humid areas tend to have acidic soil. (Like other acidic substances, acid soil is characterized by positively charged hydrogen ions, H^+.) Sandy soils also tend to be acidic; as in humid soils, water can move through them quickly.

In the western United States are arid areas where **alkaline** soils are found. These high-pH soils are associated with grass and shrubs.

Acidity or alkalinity can be stated quantitatively as a pH value. On the pH scale, which ranges from 1 to 14, 7 is neutral (the concentrations of H^+ and OH^- ions are equal). The lower the pH value of a substance is, the more acidic it is (the higher the concentration of H^+ ions); the closer to 14, the more alkaline, or basic (the higher the concentration of OH^- ions). Soil acidity depends partly on the H^+ ions in the soil water, and partly on those associated with soil particles. Some plant **nutrients** react with H^+ ions, so the availability of nutrients for plant growth is partly dependent on soil acidity.

Water

Spaces between the particles are filled with air and water. When there are large spaces between the soil particles, as in sandy soil, water and air can pass freely through the spaces. Water and air are retained more in soils that are higher in silt and even more so in clay. Most plant roots and the microbes that live in soil need both water and air.

Air

Because air in the soil evaporates from the surface and dissolves in soil water, it must be continually renewed. Many organisms—worms, insects, even small mammals—**aerate** the soil by burrowing through it. Human activities such as plowing also loosen the soil and allow aeration.

Biogeochemical Cycles

Matter and energy flow throughout the living and the nonliving parts of an ecosystem. The matter is in the form of compounds of various elements. Within an organism or a nonliving part of the environment, compounds may be broken down into their elements. Then the elements are recombined as different compounds. At each recombination, energy is used or stored in the chemical reaction. No matter or energy is lost in any change, but matter is moved and energy goes into the nonliving environment in the form of heat.

Many elements follow predictable paths through the abiotic and biotic parts of an ecosystem. These paths are called biogeochemical cycles. Carbon, for example, travels through plants, animals, air, and soil.

Carbon Cycle

Producers take in carbon dioxide (CO_2, a compound made of carbon and oxygen) from the air for manufacturing food. As a result of the food-making, the carbon atoms become part of glucose compounds; they may then become part of other compounds. When plants are eaten by animals, the carbon atoms become part of compounds in the animals and are passed along food chains through several orders of consumers.

Figure 1.4
The Carbon Cycle.

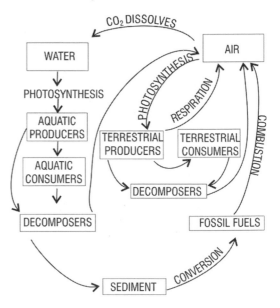

Plants and animals that die are broken down by decomposers, and their atoms are returned to the environment. At various points in that basic cycle, the carbon may take other paths, as shown in Figure 1.4. So a particular carbon atom can undergo different cycles at different times and can move from one ecosystem to another.

Water Cycle

In the water cycle, hydrogen and oxygen travel together, linked in water molecules (though each element may travel by itself when they are in other compounds). When water vapor in the atmosphere is cooled, it forms liquid water that precipitates and washes over the Earth and into streams. The water may evaporate into the air again. Or it may be used by plants and animals. Some water is eliminated from organisms, and some is passed through food chains until decomposers return it to the environment.

Figure 1.5
The Water Cycle.

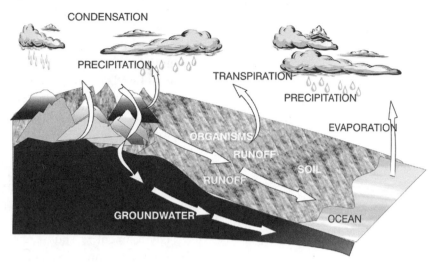

Nitrogen Cycle

The air is 78% nitrogen gas (N_2). Because plants can't use N_2, however, the atoms take an indirect route from the air to plants. Bacteria and **cyanobacteria** that live on **root nodules** of legumes convert the N_2 to ammonium ions (NH^{4+}) or nitrate ions (NO^{3-}), which can be taken up by the plants' roots and used for making protein. The conversion is called nitrogen fixation. (A small amount of nitrogen may be fixed by lightning, but it is mainly a bacterial process.)

Within a plant, nitrogen atoms become part of amino acids, the structural units making up proteins. When proteins are broken down by decomposition, the nitrogen reenters the soil or water as simpler nitrogen compounds. Soil bacteria then change them back to gaseous N_2 (denitrification).

Other Cycles

Many other elements cycle through the living and nonliving world. Among these elements are calcium and phosphorus. Although all living organisms affect biogeochemical cycles, humans do so to a greater degree. For that reason, biogeochemical cycles will be considered again in Chapter 6, Humans in the Biosphere.

The Need for Energy

All matter tends to become more disorganized, according to the physical principle called entropy. Maintaining organized living systems requires continual energy; because energy is not recycled, living systems are completely dependent on infusions of new energy for making food and carrying out activities. Ultimately, all that energy comes from the sun, in the form of photons (small packets of energy). It is absorbed by producers and passed on in one direction through consumers and decomposers, with some being lost as heat at each step. Entropy continues as organisms die and decay.

The wavelengths of light most absorbed by plants are in the blue and red

**Figure 1.6
The Nitrogen Cycle.**

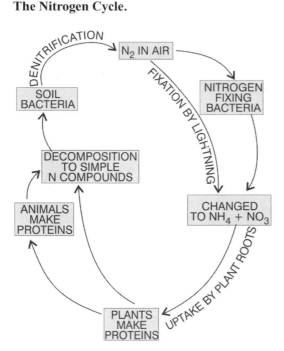

parts of the light spectrum. Least absorbed is light in the green part of the spectrum, which is responsible for plants appearing green to us.

Photosynthesis

Only green plants and some microorganisms can photosynthesize food—manufacture food from raw materials—using the energy in sunlight. They take in carbon dioxide from air that enters small openings in leaves and absorb water from the soil. The two compounds react in the important photosynthesis equation.

$$6CO_2 \text{ (carbon dioxide)} + 6H_2O \text{ (water)} \rightarrow C_6H_{12}O_6 \text{ (glucose)} + 6O_2 \text{ (oxygen)}$$

This can be read as six molecules of carbon dioxide and six molecules of water combine in the presence of sunlight to form one molecule of glucose and six molecules of oxygen. (Actually, glucose is only one compound that is formed during photosynthesis, but this standard equation is a useful overview of the process.) The energy from the sunlight is then stored in high-energy **chemical bonds** in compounds in the plant. Later the energy can be released as needed.

Like all other living things, plants are made of cells, the basic units of life. Plant cells contain structures called chloroplasts, in which photosynthesis takes place. The process combines molecules of carbon dioxide and water into complex, energy-rich organic compounds and releases oxygen to the environment. Thus photosynthesis makes a vital connection between the sun and the energy needs of living systems.

The carbon-containing (organic) molecules made in photosynthesis can be used in combination with other elements to assemble larger molecules with biological activity (including proteins, DNA, sugars, and fats). In addition, organisms can use the chemical energy stored in bonds between the atoms as sources of energy for their life processes.

Photosynthesis has two main steps, called the light reactions and the **Calvin cycle.** During the light reactions, chlorophylls and other pigments in the chloroplasts absorb photons in sunlight and convert them to chemical energy stored in the bonds of molecules.

The photons (energy-rich electrons) absorbed from sunlight by pigments are passed along a series of molecules called the electron transport system. As these electrons move along, they are replaced by electrons removed from water molecules in the plant. Water is broken down to protons, electrons, and oxygen; the protons and electrons combine with a hydrogen carrier called NADP+ (nicotinamide adenosine dinucleotide phosphate) to form NADPH. Some protons move through the membranes of the chloroplast, and their energy is used to synthesize ATP (adenosine triphosphate).

NADPH and ATP are temporary compounds that enter the Calvin cycle of reactions, in which they make three-carbon sugars from CO_2. The chemical energy is stored in the three-carbon sugars.

Chemosynthesis

The food chains near fissures in the ocean floor, from which hot water rises, are based on the energy from heat rather than sunlight (the energy source in photosynthesis). The hot water undergoes various chemical reactions with rocks. One reaction changes sulfate ions (SO_4^{2-}) in the sea water to sulfide ions (S^{2-}). Chemosynthetic bacteria then obtain energy by changing the sulfide back to the sulfate. With that energy, they manufacture food from carbon dioxide. The chemical reaction for chemosynthesis is the same as that for photosynthesis, only the source of energy is different.

Just as photosynthetic plants are the first link in most food chains, chemosynthetic bacteria are the first link in ocean-floor food chains. Animals feeding on the bacteria include giant clams, mussels, and tube worms. Crabs are secondary consumers in these communities.

Respiration

Respiration, the use of food for energy, occurs in all organisms. The chemical bonds of food molecules formed by photosynthesis (or passed along a food chain) contain energy that is released when the bonds are broken and new compounds with lower-energy bonds are formed. ATP is important in respiration, just as it is in photosynthesis: The released energy is stored in ATP's phosphate bonds. The plant or animal can later use both the glucose and the stored energy.

Glucose has six carbon atoms in each molecule. Before the organism can use glucose, it breaks down the molecule to two three-carbon molecules by a process called glycolysis. The three-carbon molecules are named pyruvate. They are changed to the compound acetyl CoA by losing another carbon atom, in the form of carbon dioxide. The acetyl CoA (which has only two carbons) then enters a cycle of reactions called the TCA (tricarboxylic acid) cycle (sometimes called the citric acid cycle or the Krebs cycle). In those reactions, the acetyl CoA combines with a four-carbon acid (oxalacetate) to form a six-carbon acid (citrate), and CoA is released. Several steps follow, with two carbons being lost as carbon dioxide. Eventually oxalacetate is formed, and that compound can combine with more incoming acetyl CoA to start a new cycle.

In addition to forming carbon dioxide, each turn of the TCA cycle stores energy in the phosphate bonds of the three-phosphate molecule ATP, which can later be broken down for releasing energy. ATP provides energy for muscular contraction and for chemical reactions in the body. When it is broken down, it becomes the two-phosphate molecule ADP (adenosine diphosphate); ADP can become ATP again through TCA cycles.

The overall reaction using one molecule of glucose is

$$C_6H_{12}O_6 + 6O_2 \rightarrow 6CO_2 + 6H_2O$$

Notice that this reaction, called respiration, is the reverse of the photosynthesis reaction. Many of the great events in the biotic world depend on these two reactions. It is easy to make the mistake of thinking one takes place just in plants and the other just in animals: Only plants can carry on photosynthesis, but *all* organisms must respire, or break down food for matter and energy.

Figure 1.7
Glycolysis and Respiration.

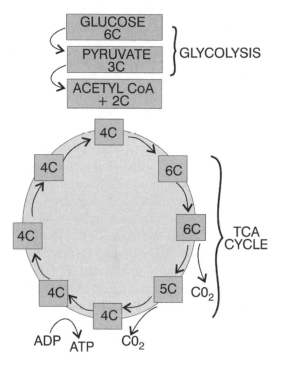

Figure 1.8
A typical population grows until it reaches the environment's carrying capacity and then fluctuates around that level.

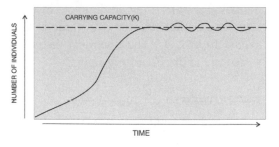

CHANGING POPULATIONS

Populations undergo continual change. They may grow larger or shrink in number; some individuals are born, die, emigrate to new environments, or immigrate from other areas. They interact with other populations in the community in predator–prey or other relationships.

Growth

Population size refers to the number of individuals in a population. Four things determine population size: birth rate, death rate, immigration rate, and emigration rate. For example, suppose a town in the Midwest has a population of 10,000 people. If 300 babies are born annually, the birth rate is 300/10,000, or 0.03. If 200 people die annually, the death rate is 200/10,000, or 0.02. One hundred immigrants arrive each year, at a rate of 100/10,000, or 0.01. However, the bored residents are leaving at the rate of 400 per year, or 0.04. At the end of a year, the population size has decreased by 200 (300 + 100 − 200 − 400).

Growth of biological populations occurs in the pattern shown in Figure 1.8. The number of individuals increases slowly at first, then rises quickly until it reaches the maximum number that can be supported by the environment. That number (symbolized by K) is the carrying capacity. Because the growth curve is S-shaped, it is called sigmoid. (The Greek letter for s is σ, sigma.) This pattern is called exponential growth; if it is graphed with the exponents of size on the y axis, a straight line will result.

Knowing the size of a population gives little information about how crowded it is or how much it will affect the environment. It is also important to know how much space the population occupies. The population size divided by the area is the population's density. At a low density, a population in a favorable environment has enough food and other resources. At a high density, it may eat too much of its supply or produce too many pollutants.

Dispersal

Animal populations may disperse (move to new areas) by flying, swimming, or walking. Plants are more limited, but they can produce seeds or **spores** that are borne by wind, water, or animals to areas that may be far from the parent plants.

Interacting Populations

Populations in the same area interact in various ways. For instance, populations coming into contact with each other may need to compete for the same resources. If one population takes more of a food source than its competitor does, then the competitor's population size will probably decrease.

Over the past few decades, coqui and greenhouse frogs that originated in the Caribbean have invaded Hawaii after being taken there in potted plants. Because the islands that make up Hawaii have no native frogs, no natural controls have evolved to check their growth, and the invaders have quickly dispersed throughout the area, reproducing and growing to huge populations. Today the frogs' croaking is so loud that many people can't sleep at night. One desperate woman set fire to her entire back yard to get rid of the frogs—which survived. Biologists are having some success in killing them with a spray that contains caffeine, but the frogs are still a problem.[3]

Competition

Living organisms have the capacity to produce populations of infinite size, but the environment and resources are finite. This fundamental tension has profound effects on the interactions between organisms.

The distribution and abundance of organisms and populations in ecosystems are limited by the availability of matter and energy and the ability of the ecosystem to recycle materials.

If Hawaii had native frogs, the Caribbean invaders would not be so successful; the natives would compete with them for food and other resources. As it is, the Caribbean frogs have moved into an empty niche and exploited it.

More than 60 years ago, G. F. Gause, a Russian ecologist, posed the principle of competitive exclusion. According to Gause, no two species could share the same niche at the same time. This is because if two species are competing for a limited resource, one species will be able to use it more efficiently than the other and will eventually crowd the other species out of the area. Which species wins, however, depends on many conditions. Also, species that may appear to be sharing a resource may not be. One example of that was shown by American ecologist Robert MacArthur, who studied five species of warblers. The five species all appeared to be competing for the same insect food in the same trees, but MacArthur discovered that members of each species spent most of their time in different parts of the trees. One fed on insects near the ends of the branches, another stayed high in the trees, a third ate insects on lower branches, and so on. In other words, each species had its own niche in the tree.

In the interspecific struggle for survival, sea otters may eat some of the fish that humans also use as food. Here, different species are competing for a resource. In other cases, different members of the same species may compete (intraspecific competition).

Commensalism

In commensalism, one species benefits and the other is neither helped nor harmed. One example is the growth of **epiphytes,** such as Spanish moss, on the branches of other plants; another is the small tropical fishes that feed on **detritus** from sea anemones while swimming between the anemones' stinging tentacles. Other fish touching those tentacles are paralyzed. The **saprophytes** that break down dead organisms are also commensals.

Parasitism

A tapeworm or leech may fasten itself to a human intestine or skin, then absorb its nutrition from the human. Unlike commensalism, parasitism benefits one species only. The **parasite** depends on the **host,** which is harmed. Usually the parasite stops short of killing the host, because that would be killing its own food source. The host may be weakened but is unlikely to die.

Mutualism

Certain species live together happily, both benefiting from the relationship; the lichens you may see growing on rocks, for instance, are actually communities of **algae,** which can photosynthesize food, and **fungi,** which can provide a **substrate** for the algae but can't manufacture their own food. They are interdependent: The algae need the moist surface provided by the fungi, and the fungi need the food made by the algae. This is known as mutualism.

Predator–Prey Interactions

These are possibly the most obvious interactions among organisms. Usually predator species do not kill off their prey species completely, but they do keep the prey populations from growing larger indefinitely.

Population Cycles

As one population is preyed on by another or is otherwise affected by it, the population sizes of both populations may

Figure 1.9
Population cycles of predators and their prey may be related, as in this example.

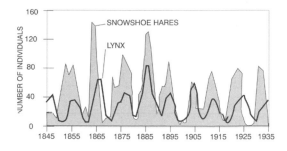

change. For example, Charles Elton (1900–1991) showed in the 1920s that fluctuations in numbers of small herbivores (such as mice, hares, and lemmings) brought about similar fluctuations in numbers of their predators, such as the arctic fox. Wolves may lower the size of a population of deer, and the lowering of the deer population can lessen the damage to trees on which the deer feed. Eventually, depending on other environmental factors, the plants may increase so much that the deer population increases along with their food supply. Or there may be so few deer that the wolves can't find enough prey, and the number of wolves then declines. The rises and falls of populations over the years are known as population cycles.

After some event such as a forest fire or volcanic eruption disturbs a landscape, a series of plant communities—first herbs, then shrubs, and finally trees—move in to establish a new community. American ecologist Frederic Clements (1874–1945), who first pointed out this principle, thought that for any area there was a rigid order of succession ending with a climax community. Biologists have since shown that the order of succession, as well as the final community, is not entirely predictable.

Change over Time

Life on Earth today is a rich, mosaic tapestry of species interacting in many ways and forming large patterns in the biosphere. As Ricketts and Steinbeck realized, by studying individual organisms and their connections, one can begin to grasp the whole of nature, "from the tidepool to the stars."

Yet even the fertile picture provided by ecology is flat and two-dimensional without a knowledge of life's history. Where did each species come from? How did it form its present connections with other organisms and with the nonliving environment?

Questions like these lead to the field of evolution, which provides the necessary third dimension to the ecological picture. Evolution will be discussed in Chapter 2.

NOTES

1. J. Steinbeck and E. Ricketts, *Sea of Cortez* (Mount Vernon, N.Y.: Paul P. Appel, 1941), pp. 256–57.

2. F. Clements, 1905; cited in V. E. Shelford, *The Ecology of North America* (Urbana: University of Illinois Press, 1963).

3. From "Living on Earth," broadcast on National Public Radio, January 5, 2001; used by permission.

The Evolution of Life

The Ecuadorian cargo ship *Jessica* ran aground near some Pacific islands on January 16, 2001, and within a few days diesel and bunker fuel began spilling into the water. Eventually 160,000 gallons of the oily stuff spread throughout the area, threatening to poison huge numbers of animals and perhaps to damage the entire ecosystem. All over the world, environmentalists were horrified by the oil spill and its possibly disastrous results.

Luckily, only a few dozen birds and sea lions were poisoned; as often happens with oceanic spills, westward winds and ocean currents carried the spill away within a couple of weeks, lessening its impact on the islands. Ecologists were greatly relieved but still apprehensive about future spills.[1]

Why were ecologists so upset about this oil spill, when many other spills have caused less outrage? It was because the threatened islands were the Galápagos, the home of blue-footed boobies, unique tortoises, iguanas, and other animals studied by Charles Darwin (1809–1882), the man most people associate with the idea of evolution. The islands were an irreplaceable natural record of Darwin's evidence.

Darwin's **theory** of **evolution** by **natural selection** was published—as *The Origin of Species by Natural Selection, or the Preservation of Favoured Races in the Struggle for Life*—in 1859 and revolutionized the popular idea of evolution. However, the idea of evolution itself had been accepted by many biologists long before Darwin; what was missing was a theory of *how* organisms changed over time. Darwin provided that theory.

EXPLAINING CHANGES OVER TIME

Ancient Greek scholars had ideas related to changes in plants and animals; Empedocles (c. 490–430 B.C.) was one of the first to suggest that the entire universe was made up of four elements (earth, water, fire, and air). Empedocles thought that all living things were the product of "evolution" of separate body parts. Anaximander (610–c. 545 B.C.) believed that life originated in the sea and that all land animals are descended from sea animals. His successor, Anaximenes of Miletus (flourished c. 545 B.C.), thought that air (rather than water) was the origin of all things. Although these ideas are not part of the modern theory of evolution, they did contain the seeds of the concept that living things can change over time.

Greek philosopher and scientist Aristotle (384–322 B.C.) did the first early work on classification, describing about 500 kinds of animals. He classified animals in a ladderlike *Scala Naturae* or "Chain of Being." The ladder's "rungs" included, in descending order, God, man, mammals, oviparous with perfect eggs (such as birds), oviparous with nonperfect eggs (such as fish), insects, plants, and nonliving matter. All of these groups were assumed to be immutable.

As in most religions, the Judaic and Christian Bibles contained a creation story about each kind of living thing being created separately and the account was taken literally. Because Christian theology dominated European intellectual life in the Middle Ages, any suggestion that one species could evolve into another was considered heresy. Although biologists made advances in medicine and natural history during those centuries, the Christian view of separate creation was generally accepted.

The Age of Discovery

In the sixteenth and seventeenth centuries, explorers from western Europe traveled to parts of the world that had been unknown to them, and discovered strange new plants and animals. In the Americas they found what English physician and essayist Thomas Browne (1605–1682) called "Beasts of prey and noxious Animals," but no horses. Browne puzzled over how all those unknown animals had made their way from Noah's ark on Mount Ararat in Turkey to the New World without leaving any descendants along their path through Europe. Speculation such as Browne's

helped pave the way for acceptance of the idea that organisms have changed over time and that evolution can lead to different organisms in different parts of the world.

Browne accepted the Aristotelian idea of the *Scala Naturae,* modified a bit for Christianity. At its bottom were minerals; at higher steps were plants and "lower" animals; near the top were human beings; and at the pinnacle were angels and God. Each group was considered to be just like it had been at the creation, and there was no thought of the separate groups being related to each other.

Other Englishmen—natural historians like Gilbert White (1720–1795) and John Ray (1627–1705)—accumulated detailed descriptions of plants and animals that prepared readers for Darwin's idea that the variation in natural populations is one of the necessary conditions for evolution. They also attempted to classify the organisms they studied on the basis of similarities in structure.

Carl Linnaeus (1707–1778), the Swedish father of taxonomy (see Chapter 4), contributed a system of classification that is still used in modified form today, though he believed in special creation.

Preludes to Darwin

These English and Swedish thinkers contributed to the changing ideas about fixed species of organisms, but French philosophers went further. Count Georges-Louis de Buffon (1707–1788) wrote about the struggle for existence, the tendency for life to multiply at a faster rate than its food supply, variations within species, similarities of structure in different organisms, a very long time scale, the extinction of some earlier animals, geographic distribution of plants and animals, and other matters that formed important parts of Darwin's 1859 theory. Darwin wrote of them in the following century and became famous; Buffon is remembered only by historians of science because he never brought all his ideas together in a single book or proposed a mechanism for change. Like many others, he came close to Darwin's achievement but failed to provide a new **paradigm.**

Back in England, Erasmus Darwin (1731–1802), Charles's grandfather, observed various adaptations, such as protective coloration. He wrote about ecological relationships and about living **fossils** possibly surviving at great depths in the ocean. He proposed the idea of **sexual selection** and said that all plants and animals are continually changing. But he thought erroneously that organisms acquired new characteristics and then passed them on to their offspring—a process called the inheritance of acquired characteristics.

The inheritance of acquired characteristics is usually associated more with French biologist Jean-Baptiste Lamarck (1744–1829), who thought animals strove to better themselves. Probably he and Erasmus Darwin reached their theories independently; they were simply working in the same intellectual climate.[2] Like many of their predecessors, both scientists thought organisms unconsciously **adapted** to their environments and passed on their adaptations to their offspring.

Geology

The next important intellectual step leading to Darwin's theory was the idea of geologic time, measured in hundreds of millions of years. Even if scientists accepted the idea of evolution, it would have been hard to imagine modern plants and animals evolving from very simple forms in the few thousand years implied by the Bible. James Hutton (1726–1797), the founder of historical geology, first proposed that the Earth's crust alternately was uplifted and subsided because of dynamic forces acting within it during geologic time periods. Hutton also distinguished between **igneous rock** and **sedimentary rock.** His ideas made it possible to imagine seabeds (with their enclosed fossils) becoming mountain tops or mountains being covered by oceans.

In 1815 William Smith (1769–1839) published a map of the British Isles that was like no other prepared before that time. Smith was a surveyor and canal digger who later became lauded as the father of geology. Walking hundreds of miles and working all alone, Smith had collected data on rock outcrops and fossils that he used for drawing the map. The map showed the hidden layers of rock beneath England's surface, but Smith also realized that the fossils in different layers were specific to those layers. He understood that by following the fossil trail he could trace rock layers as they rose and fell across England and even around the world.

During the nineteenth century, many fossils of prehistoric plants and animals were uncovered. Dinosaur skeletons, as well as the bones of early humans, were unearthed and reassembled, and special museums were built to house the skeletons. Baron Georges Cuvier (1769–1832), a French paleontologist, used his knowledge of comparative anatomy of living animals to make celebrated reconstructions of fossil animals from only a few of the fossil bones. He also dissented from the prevailing Chain of Being theory, which showed a single path from the simplest to the most complex organisms. Cuvier realized there were sharply different animal groups that could not be fitted into that path. However, Cuvier rejected the idea of evolution and proposed a theory of catastrophism. He thought that each of a series of catastrophes had made old forms extinct and allowed the creation of new life.

Charles Lyell (1797–1875) was only 11 years older than Darwin; his *Principles of Geology* was of major importance for Darwin's theorizing. In fact, it contains much of what Darwin would write in *The Origin of Species*—with the important exception of natural selection as the mechanism of evolution. Again, a scientist came close to Darwin's achievement but failed to spot the crucial point.

Fossils

The great overall picture of evolution that you might see in a museum, based on the evolution of species, is called macroevolution. It shows how entire organisms have changed over time. Much of that picture is based on skeletons and other fossils.

The evidence from fossils—bones, leaves, shells, and other recognizable evidence of past life—has been crucial in leading people to ask questions about where the organisms of today came from and to find some of the answers. Scientists who study and interpret fossils are called paleontologists.

Though fossils may be found in rock layers in many parts of the world, some areas are more informative than others. The Grand Canyon in northern Arizona, for instance, was created by the Colorado River gradually carving out a channel through rocks aged as much as 1.7 billion years, along with their enclosed fossils. The rocks surrounding the fossils are arranged in easily distinguishable colored layers referred to as pink cliffs, gray cliffs, vermilion cliffs, and so on. Some kinds of fossils are found only in specific layers.

When layers of rock are built up gradually by the settling of mud, sand, or volcanic ash, fossils may be trapped in the rock layers. Some settle to the bottom of drying lakes, for instance. Others may be carried by moving water to spots where they settle and are covered over by the **silt** and sand in the river or stream that carried them. Much later, more fossils are deposited in rock layers above the first one. This process has continued for about 3.8 billion years. The forces that compress the rock layers are great, so a **geologic era** that lasted for millions of years may be shown as a rock layer only a few millimeters thick. Rocks formed by the accumulation of particles of gravel, sand, and mud are called sedimentary rocks. With the passage of time and the accumulation of more particles, and perhaps with chemical changes as well, various types of sedimentary rock—conglomerate, sandstone, mudstone, or shale—are formed.

In an undisturbed series of rock layers (strata), the oldest layers are, of course, the bottom layers, and the youngest are at the top. However, earthquakes and other movements of the Earth's crust can disturb these strata, tilting them sideways. Some layers are turned to an almost vertical position. For that reason, in inferring the ages of the strata and the fossils found in them, paleontologists must use caution. They may use **radioactive dating** methods and compare results from many different areas.

In thinking of fossils, people are likely to think first of dinosaurs, but those animals were only a small fraction of the millions of species that have ever lived on Earth. Most of the fossil record consists of fossils of shelled animals and other **invertebrates.**

In the late eighteenth and early nineteenth centuries, English and French paleontologists and engineers—notably William Smith—discovered that even when rocks of the same age are separated by great geographic differences, they may contain the same fossils. (Geologists refer to fossils of organisms that lived for a short, well-known time on the geologic scale as index fossils, because they indicate the ages of rocks.) By observing rocks and their fossils, the men were able to recognize rocks of the same age that were on opposite sides of the English Channel.

The kinds of fossils in rocks of different ages are different, because Earth has changed and the animals and plants have changed as well. Most of the fossils found are the remains of extinct organisms that no longer live anywhere on the planet. In fact, more than 99.9% of the animal species that have ever lived on Earth are now extinct.

Index fossils are very useful for geologists, but a more accurate way of measuring the age of rock layers is now available. Because the minerals in rocks contain naturally occurring radioactive elements that decay at constant rates over time, paleontologists can measure the proportion of decayed elements in a rock and compute its age.

In some cases, fossils have been faked by pranksters or by people who want to discredit the very idea of gradual evolution. In 1908, the first bones were unearthed of what would become the most famous scientific fraud in history—Piltdown Man, or *Eoanthropus dawsoni.* In 1912 the Piltdown skull was discovered in England. Amazingly, the skull combined a large human cranium with an ape's jaw. Scientists were astounded at the find but became convinced that it supported the argument that humans developed large brains before they walked upright. (We know now from much more evidence that human evolution proceeded in just the opposite order; our ancestors walked upright while they still had small, apelike brains.) Eventually the fraud was revealed: Someone had put together a modern human skull with the jawbone of an orangutan, with its teeth filed to look as if they were human. Probably it will never be known with certainty who perpetrated this stunt or why. Biologist Stephen Jay Gould (1941–2002) suggested that the young anthropologists involved in the discovery were playing an elaborate practical joke that got out of hand. Soon afterward one of them (Charles Dawson, 1864–1916) was killed in World War I, and another (Pierre Teilhard de Chardin [1881–1955], later a revered professional anthropologist and Catholic theologian) may have been unable to back out of the situation.[3]

In 1999, someone offered for sale an interesting new fossil named *Archaeoraptor,* which appeared to be a dinosaur with feathers. Private fossil collectors or museums could be expected to pay a great deal for this fossil, which seemed to add to the accumulating evidence that modern birds descended from dinosaurs. Paleontologists were skeptical about the find, however, and within a few months it was unmasked as a fake. Someone had added a reptilian tail to a real feathered skeleton in the hope of making it more valuable as a collector's item; in the process, they damaged the real fossil, which was immensely valuable as scientific evidence.

Theorizing

Fakes like Piltdown and *Archaeoraptor* are only temporary hindrances to the growth of knowledge about evolution, and they can seem amusing. Unfortunately, they are also used as ammunition against the whole idea of gradual evolution, used by **creationists** who believe that the biblical account of creation is literally true. Most biologists believe that the Genesis version of creation is a **myth** or **allegory,** not a description of actual events. Since the debunking of *Archaeoraptor,* some creationists have claimed that the forgery is evidence against evolution. They fail to understand that science thrives on controversy, it is self-correcting, and it advances partly by disproving mistaken ideas.

One twentieth-century philosopher of science, Karl Popper (1902–1994), said that science approaches the truth by the falsification of hypotheses. This process is based on the logic taught by Aristotle. According to Popper, a scientist gathers observations about the natural world, then forms a tentative hypothesis to explain the observations. Further observations or experiments may support the hypothesis indefinitely; but if even one reliable observation or controlled experiment falsifies the hypothesis, it must be rejected or revised. Popper's philosophy is not widely accepted in the scientific community today, but still guides some research. Although it may seem like a negative approach to discovering the truth, it is more like a process of gradual refinement. As fictional detective Sherlock Holmes once said, "It is an old maxim of mine that when you have excluded the impossible, whatever remains, however improbable, must be the truth."[4]

In biology, it is difficult to design falsifiable experiments that can test the whole world of organisms. It is often more likely that a scientist will gather more observations, try to comprehend an overarching principle to account for them, and propose a theory explaining the principle. Later scientists can test their own observations against that theory.

Charles Darwin

Darwin may be history's greatest example of a scientist gathering data and formulating a theory gradually. Though his famous book about the theory of natural selection appeared in 1859, he published it only after 20 years of study and reflection, along with much personal anguish about how it would be received.

Darwin's personal voyage to his theory began when he was a young man, the younger son of an English physician and grandson of noted biologist Erasmus Darwin. The young Charles was unsure of what he wanted to do with his life, though he was interested in botany and geology because of some college classes he had taken. By chance, at the age of 22 he was offered a job as a ship's naturalist, on the H.M.S. *Beagle.* In 1831 the *Beagle* was going to set off on an around-the-world voyage, and the captain needed to have a naturalist sketch the animals and plants that would be seen in far-off lands. Charles jumped at the chance for adventure, little realizing how he—and the world—would be changed by it.

For five years after embarking from London, the *Beagle* sailed from one country or island to another. It sailed down along the coast of Europe, across the Atlantic Ocean to South America and around the Cape to the Pacific. It crossed the Pacific and Indian Oceans, steered south and around Africa's Cape of Good Hope, and back to the Atlantic. Throughout the voyage, Darwin took careful notes, made drawings in his journal, and collected both living specimens and fossil bones.

In the long hours aboard the ship, Darwin read some of the books he had taken along. One of them was *The Principles of Geology,* by geologist Charles Lyell, which had been published in 1830. In contrast to the prevailing idea—catastrophism—that major changes in the Earth have resulted from catastrophes, such as floods and volcanic eruptions, Lyell's argument was that the same natural forces—wind, water, fire, and so forth—have shaped Earth since its beginning and are still present. Given enough time, those forces could have produced both the present landscape and the record of the past seen in rock layers.

Lyell's theory greatly influenced Darwin's thinking about changes in organisms.

The Galápagos

When the *Beagle* reached the Galápagos Islands, 600 miles west of Ecuador in the Pacific Ocean, Darwin found some especially interesting animal life. The Galápagos tortoises had shells of varying shapes, with the shape on each island being unique to that island. Sailors who caught tortoises as a convenient supply of fresh meat on board told Darwin that they could tell immediately by examining the shell which island a tortoise lived on.

He collected specimens of various kinds of birds, including 14 kinds of Galápagos finches (*Geospiza* species). Later on he said:

The most curious fact is the perfect gradation in the size of the beaks in the different species of

Geospiza. . . . there are no less than six species with insensibly graduated beaks. . . . Seeing this gradation and diversity of structure in one small, intimately related group of birds, one might really fancy that from an original paucity of birds in this archipelago, one species had been taken and modified for different ends.[5]

(As it turned out, the Galápagos finches not only were important evidence in Darwin's theory but are still providing biologists with important information. Peter R. Grant and his colleagues at Princeton University studied the effects of a recent dry spell on the Galápagos finches and found that natural selection occurred quickly. The lack of water killed all the plants except those with large, drought-resistant seeds. Most of the finches are seed-eaters, so many died for lack of food. The surviving finches were those with large beaks, which could crack open the large seeds. During a series of wet and dry conditions on the

Figure 2.1
A Map of the H.M.S. *Beagle*'s Route. Of the voyage's five years, three and a half were spent in and near South America.

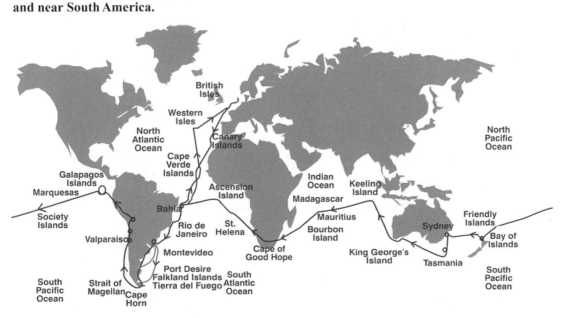

islands, the selection process occurred over and over, often within just a few months. Some biologists have objected that this is not real selection, but a normal cycle made possible by genetic variation in a population. Nevertheless, it is an excellent illustration of how selection operates.)

When Darwin returned to England, he married and settled down in a home in the country where he could review his findings and publish them. He was extremely careful, examining the specimens he had collected and reading papers and books by a variety of writers.

Malthus

In reading *An Essay on the Principles of Population,* written in 1798 by English schoolmaster Thomas Robert Malthus (1766–1834), Darwin came across a theory of economics that profoundly affected his thinking about biology. Malthus had written that although the world's supply of food can increase only a little at a time, the human population increases in "geometrical ratio." (That is, it increases by doubling, redoubling, and so on. After just a few doublings, a very high number is reached.) Malthus pointed out that the human population must be kept in check by wars, plagues, and other disasters so as not to outstrip the food supply. Darwin realized that the same principle must apply to nature in general—that there must be a continual struggle for survival among the plants and animals competing for the same food and other resources. In that struggle, those organisms best adapted to their particular environments can survive long enough to reproduce, but others die young. Later on, Darwin recalled, "It at once struck me that under these circumstances favourable variations would tend to be preserved, and unfavourable ones to be destroyed. The result would be the formation of new species." Over many generations, the heritable "favorable variations"

that led to survival of individuals would bring about **speciation,** with the new species being well adapted to its environment because of natural selection by that environment.

Even aside from his scientific hesitancy about publishing, Darwin delayed having his work published. He realized that it would cause an uproar because of the conflict with many people's religious beliefs. (His wife, Emma Wedgwood Darwin, was also a devout Christian.) The idea that species could change, and especially that humans were the result of gradual evolution from earlier beings, would be as upsetting in the Victorian world as the Copernican theory that the Earth was just one of the planets revolving around one star had been in the sixteenth century.

Wallace

Years went by, during which Darwin did research on barnacles and earthworms, attended scientific meetings, and read. He collected all the data possible to buttress his theory of evolution by natural selection and began writing the manuscript of a book about it. Yet he had not published. He became a semi-invalid, taking refuge in his home. Finally, the decision was taken out of his hands: He received a remarkable letter and essay from Alfred Russel Wallace (1823–1913), a younger biologist.

Wallace wrote to Darwin from the East Indies, where he was doing research. Darwin was shocked when he read the essay, realizing that Wallace had came up with almost the same theory of evolution. He wrote to some scientist friends (Lyell and others), asking for their advice. When the scientists saw Wallace's paper, they were taken aback, knowing that Darwin had been working on the theory for about 20 years. They prevailed on him to publish at last, so that the credit for his idea of natural selection would not go to Wallace. They also persuaded Wallace to present his own theory jointly with Darwin, establishing Darwin's priority.

Reluctantly, Darwin agreed to finish his book. As he had expected, there was a storm of protest, one that has continued for nearly 150 years. As another result, though Darwin and Wallace eventually published some papers together, Wallace never became as famous as Darwin.

A Summary of Darwin's Theory

Darwin's theory has been summarized as follows by biologist and educator R. W. Lewis (1911–2000):[6]

1. All life evolved from one simple kind of organism.
2. Each species, fossil or living, arose from another species that preceded it in time.
3. Evolutionary changes were gradual and of long duration.
4. Each species originated in a single geographic location.
5. Over long periods of time new genera, new families, new orders, new classes, and new phyla arose by a continuation of the kind of evolution that produced new species.
6. The greater the similarity between two groups of organisms, the closer their relationship and the closer in geologic time their common ancestral group.
7. Extinction of old forms (species, etc.) is a consequence of the production of new forms or of environmental change.
8. Once a species or other group has become extinct, it never reappears.
9. Evolution continues today in generally the same manner as during preceding geologic eras.

Darwin's theory of "descent with modification," as he called it, has stood the test of time. Although it has been modified a little, his general principles are still accepted. In addition, modern studies of **DNA** and other molecules have helped explain some of the things Darwin didn't understand.

Many areas of biology provide evidence supporting the theory of evolution by natural selection. Among the items of evidence, as listed by the National Academies of Science,[7] are:

Paleontology	Ancient bones and other fossils provide clues about what organisms of the past looked like and when they lived.
Anatomy	The structures of different organisms are based on similar plans, with differences reflecting their requirements for survival in different surroundings.
Biogeography	The myriad kinds of organisms on Earth, and their many specialized niches, are best explained by natural selection for useful adaptations.
Embryology	The embryos of very different animals, such as chickens and humans, look remarkably similar at certain stages in their development. Common ancestry seems to explain the similarities best.
Molecular biology	DNA is very similar in all organisms, especially in those that are closely related. The same 20 amino acids are used in building all kinds of life. Some stretches of DNA represent "excess baggage" carried from ancestors but not used in their descendants.

EXPLAINING GEOGRAPHIC SIMILARITIES AND DIFFERENCES

If you look at a modern map of the world, you can see a striking resemblance between the coastlines of South America and Africa; in fact, they can be fitted together almost like neighboring pieces of a jigsaw puzzle. Perhaps even more striking is the resemblance between fossils of many species on the two continents; though parts of Brazil and South Africa are about 2,418 km (3,900 miles) apart, many plants and animals in the two areas, as well as their layers of rock, are much alike. Could this just be an interesting coincidence? Interdisciplinary scientist Alfred Wegener (1880–1930) didn't think so. In 1915 he published a theory that was ridiculed by many scientists—the theory of continental drift. Wegener proposed that all the modern continents were at one time united in one supercontinent he named Pangaea, and it gradually broke up into separate continents. A northern continent called Laurasia later separated farther to become North America, Europe, and Asia; a southern continent, Gondwanaland, gave rise to South America, Antarctica, India, and Africa.

In the late 1960s, a modified form of Wegener's theory finally became accepted by scientists, because new evidence—especially new maps of the sea floors—supported it rather than falsifying it. Today it has been subsumed by the general idea of **plate tectonics,** one of the most important theories in geology.

Wegener believed that the continents simply moved through the oceans. Today, scientists know that continents move because they are the higher portions of huge crustal plates that float on a hot, semiliquid rock layer called the asthenosphere. (Oceans are the submerged portions of the plates.) Where continental plates are moving away from each other, ocean ridges are

Figure 2.2
According to Wegener, the continents were joined in one large mass (which he called Pangaea) during the Carboniferous era. These later moved apart, finally reaching their present positions. In this picture, the black areas are oceans; the gray areas, shallow seas; and the white areas, land masses.

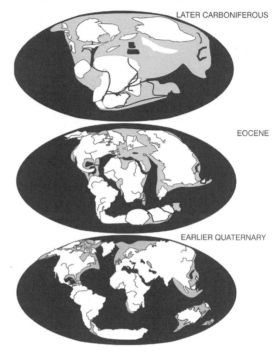

found on the ocean floor; here molten rock rises from below the crust, flows outward, and hardens to become new crust. Where plates collide with each other, they may push up mountains, such as the Himalayas. In some places, one plate may slide under another, creating an oceanic trench and chain of volcanoes. This revision of Wegener's theory is called plate tectonics.

Movements of the crustal plates are still continuing, causing earthquakes and geographic movements. The plate containing southern California, for instance, is slowly moving northward, and in about 150 million years it will collide with Alaska.

Figure 2.3
Land masses have reached their present positions by riding on tectonic plates, which are still moving.

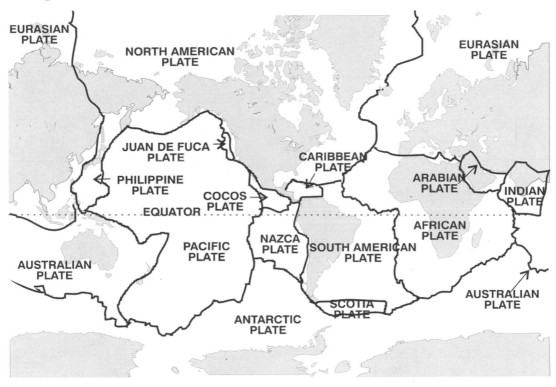

When the continents moved apart, the organisms on them were separated. Many animals and plants on Gondwanaland moved to South America and Africa. Fossil evidence supports this, showing similar crocodilians and dinosaurs on both continents during the early Cretaceous period. Once separated, the South American and African relatives could evolve in separate directions.

From various studies during the nineteenth and twentieth centuries, scientists constructed the geologic eras shown in Figure 2.4.

The Major Eras of Life

Reading matter about paleontology and geology contains references to a great many different names of periods and rock layers in the history of life on Earth. In this book, we restrict ourselves to the major eras and their features. Other terms you may encounter can be put into proper perspective by their ages.

For many years, the oldest fossils known were those of the Paleozoic era, which began 544 million years ago. (The name comes from the Greek words *paleos,* which means old, and *zoos,* meaning life.) Because the oldest part of the Paleozoic was called the Cambrian period, everything before the Paleozoic was called the Precambrian era. Eventually scientists discovered fossils in Precambrian rock **strata,** and that period was subdivided into other epochs.

After the Paleozoic era came the Mesozoic era, beginning 245 million years ago.

Figure 2.4
Geologic Time.

TIME (of beginning)	ERA	PERIOD	EPOCH	MAIN EVENTS
4500 MYA	Hadean			Formation of Earth
3800 MYA	Precambrian		Archaean	Earliest life: algae, bacteria
2500 MYA			Proterozoic	Early marine animals
544 MYA	Paleozoic	Cambrian		Marine animals
505 MYA		Ordovician		Early fishes
440 MYA		Silurian		Land plants and animals
410 MYA		Devonian		Amphibians
360 MYA		Carboniferous		Amphibians, reptiles Insects, club mosses Horsetails
286 MYA		Permian		Mammal-like reptiles
245 MYA	Mesozoic	Triassic		Early mammals, cycads Conifers, dinosaurs
208 MYA		Jurassic		Peak of dinosaurs Birds
146 MYA		Cretaceous		Flowering plants, bees
65 MYA	Cenozoic	Tertiary	Paleocene	Mammals, Extinction of dinosaurs
54 MYA			Eocene	Early horses
38 MYA			Oligocene	Grasses, grazing mammals
23 MYA			Miocene	Increase in mammals
5 MYA			Pliocene	Later hominids
1.8 MYA		Quaternary	Pleistocene	*Homo sapiens*
11,000 (not in millions)			Holocene	Human civilizations

36

Figure 2.5
Relative lengths of geologic eras.

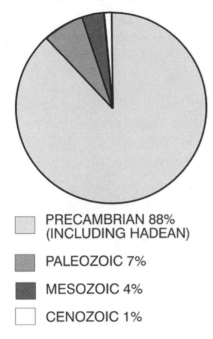

PRECAMBRIAN 88%
(INCLUDING HADEAN)

PALEOZOIC 7%

MESOZOIC 4%

CENOZOIC 1%

Its name comes from the Greek word *mesos,* meaning middle.

The last era, the Cenozoic era, began 65 million years ago and continues today. Its name is based on the Greek word *coenos* (recent). Some tables of geologic eras show the Cenozoic, Mesozoic, and Paleozoic eras as divisions of the Phanerozoic Eon.

Hadean Time

The stage was set for life on Earth in the mysterious time when the planet was being shaped from a conglomeration of gas and dust 4.5 to 3.8 billion years ago, when our solar system was forming. The period is sometimes called Hadean time.

As our planet and others were being formed, collisions of coalescing particles released so much heat that the early planets were molten, not solid. As the planets took shape and collisions among the particles less-ened, the planets cooled. The molten material forming the surface of Earth first solidified into rock about 3.8 billion years ago.

When rock itself appeared, Earth's geological history could begin. It was the beginning of plate tectonics, of the appearance of mountains and their erosion.

Precambrian

For more than a billion years more, Earth was lifeless. The rocky surface was still cooling, and the atmosphere—containing ammonia, methane, and other gases—would have been poisonous to the organisms of today. There was as yet no oxygen.

ARCHAEAN

Some time between 3.8 and 2.5 billion years ago, the first life appeared on Earth. The oldest fossils found so far—microfossils of **archaeans** and other bacteria—have been dated as 3.5 billion years old. (For more details about the archaeans, see Chapter 5.)

For about a billion years, all life was composed of bacteria. Our evidence for this is fossils of stromatolites in South Africa and Western Australia. Stromatolites are colonies of photosynthetic bacteria; when these organisms appeared, they gave off oxygen.

PROTEROZOIC

Life expanded in the Proterozoic (2.5 billion years ago to 544 million years ago) period, the latter part of the Precambrian era. During this time, stable continents first appeared, and the first abundant bacteria and archaeans lived and left fossils. Oxygen began to build up in the atmosphere as a result of photosynthesis; thus, for the first time, aerobic bacteria could appear and thrive.

Organisms made of cells with **nuclei** made their appearance about 1.8 billion years ago. Some anaerobic bacteria died as oxygen accumulated, but aerobic organisms

Figure 2.6
The Precambrian Era.

took advantage of the changing atmosphere. Multicellular algae and finally the first animals appeared.

Paleozoic

Though the Paleozoic is often called the Age of Fishes, and fishes did dominate that era, the first land vertebrates also appeared then. In fact, one of the major events of the Paleozoic era was the emergence of some animals onto land from the water. The Earth was becoming drier, and adapting to new environments meant acquiring new structures for crawling instead of swimming and for absorbing oxygen from air instead of from water.

Long before the **tetrapods** and other large Paleozoic animals appeared, however, the Cambrian period began the Paleozoic era with what is sometimes called the "big bang" of animal evolution. About 544 million years ago, the simple one-celled organisms that had populated the Earth became multicellular. For the first time, they could grow large and could form **tissues** that were capable of carrying out separate functions within an organism. Within about 10 million years of the Cambrian period, an enormous array of multicellular animals appeared on the scene. Though many of them were destined to die out later, among them were the distant ancestors of all the animals on Earth today. The animals that emerged and radiated then included sea cucumbers and jellyfish, mollusks and lamp shells, carnivorous worms, bristle worms, roundworms, and the arthropod ancestors of spiders, insects, lobsters, and crabs.

During the first 60 million years of the Paleozoic era, all the major phyla of animals appeared—Annelida (segmented worms), Echinodermata (starfish and their relatives), Mollusca (clams, snails), Arthropoda (insects, crustaceans), Chordata (fish, amphibians, reptiles, birds, mammals), and others. In the hundreds of millions of years since then, no new phyla have appeared.

Figure 2.7
The Cambrian Period of the Paleozoic Era.

Natural selection of animals has acted on those phyla, each with its own characteristic body plan. They have evolved in many ways in response to changing environmental conditions while retaining the basic features of their phyla.

Much evidence about the Cambrian period lies in the Burgess Shale rock layer in the Canadian Rockies. However, in recent years evidence of the same explosion of life has been found in many other parts of the world. Something in the world environment or in the animals themselves must have changed quickly in geological terms.

Sometime in the Devonian period a lumpy creature that was basically a fish crawled out of the water in what is now southern Quebec. It was about 1.5 to 2 meters long, had stumplike fins, and could probably breathe air for brief periods using primitive lungs. The creature was *Eusthenopteron,* a crossopterygian (lobe-finned) fish. *Eusthenopteron* had some fea-

tures that resemble those of modern amphibians, such as salamanders, including lungs. Other characteristics, though, were more fishlike.

Anthropologist Loren Eiseley (1907–1977) once described the first emergence onto land in this dramatic fashion:

It began as such things always begin—in the ooze of unnoticed swamps, in the darkness of eclipsed moons. It began with a strangled gasping for air.

The pond was a place of reek and corruption, of fetid smells and of oxygen-starved fish breathing through laboring gills. At times the slowly contracting circle of the water left little windrows of minnows who skittered desperately to escape the sun, but who died, nevertheless . . .

On the oily surface of the pond, from time to time a snout thrust upward, took in air with a queer grunting inspiration, and swirled back to the bottom. The pond was doomed, the water was foul, and the oxygen almost gone, but the

Figure 2.8
The Devonian Period of the Paleozoic Era.

creature would not die. It could breathe air direct through a little accessory lung, and it could walk. . . . It walked rarely and under protest, but that was not surprising. The creature was a fish. . . . Though he breathed and walked primarily in order to stay in the water, he was coming ashore.[8]

In addition to spawning the amphibians, the fishes of that time evolved rapidly; some became the placoderms, or jawed fishes. Others became the cartilaginous and bony fishes. Today fishes are found in every freshwater and saltwater environment. Fishes are more abundant and varied than their air-breathing tetrapod descendants. Though confined to the water, they have created or exploited numerous niches in that environment for the past half billion years.

Later in the Devonian, true tetrapods, "fish with legs," appeared and moved onto land. During the Carboniferous period, the labyrinthodonts (early amphibians) became dominant life forms and diverged into many species.

Mesozoic

Tourists driving across Utah are likely to stop at Dinosaur National Monument, the site of a gigantic deposit of dinosaur bone fossils. Although many museums have displays of assembled dinosaur skeletons, at Dinosaur National Monument the process of excavation can be seen also, because some of the skeletons are still partly embedded in the rocks.

The variety of dinosaurs in Utah, as well as the number, is impressive. Some, such as *Ornitholestes,* were chicken-sized; some, such as the sauropods, were gigantic; others were in between. Both herbivores and carnivores are there. Some lived on land, others in swamps. They lived in the Jurassic period of the Mesozoic era, an age recorded in the rock layer called the Morrison formation.

At the beginning of the Mesozoic, small land masses had joined together and formed the supercontinent Pangaea. As the era continued, Pangaea began to break apart again.

Figure 2.9
The Mesozoic Era.

Laurasia and Gondwana began to divide by the Middle Jurassic period, separated by the Sea of Tethys. Later, the part of Laurasia that is now North America began to move away from the rest of Laurasia and Gondwana. Australia and Antarctica pulled away from India by the later Jurassic, and South America and Africa began to separate.

Ecosystems in the Mesozoic were affected by the shallow waters that moved over the land and back to the seas. This created warm, wet environments on land. In the Cretaceous period, large areas of all the continents were flooded for the last time. One sea covered North America from the Gulf of Mexico to the Arctic.

These conditions created lowland swamps, but mountains were being built as well. The mountain ranges we see today along the western edges of North and South America appeared, as did other mountains in separated parts of Gondwana.

At the end of the Paleozoic era was a major extinction, known as the Permo-Triassic crisis. This catastrophe affected most of the species living then, whether they lived in shallow waters, in the deepest parts of the sea, or at the margins of continents. Crinoids ("sea lilies"), **bryozoans, brachiopods,** and other groups disappeared. Only a few ammonites survived, but they proceeded to multiply and dominate the Mesozoic seas. Plant species also declined greatly, as did the mammal-like reptiles. It was the greatest crisis in the history of life on Earth.

New niches were available in the aftermath of that extinction, and the first true mammals—small shrewlike omnivores—appeared and filled some of the niches. Also, the reptilian ancestors of lizards, crocodiles, and turtles evolved. The Triassic period land environments were dominated, however, by the two major reptile groups: therapsids, or mammal-like reptiles, and thecodonts, which were the ancestors of the dinosaurs. Then the Triassic ended in another extinction that wiped out about 35% of all animal groups, 80% of the reptiles, and many of the ammonites.

The next period, the Jurassic, is famous for the dinosaurs that dominated Earth's land, sea, and sky for more than 100 million years. (Human beings, in contrast, have lived on the Earth less than one-tenth as long.) Though the dinosaurs themselves became extinct at the end of the Mesozoic, some of them (the theropods) gave rise to the birds.

A young dinosaur less than a meter long provides the best example of dinosaurs that were probably ancestral to birds. Named *Bambiraptor,* it was discovered in Montana in 1994 by a teenaged fossil hunter. It had evolved structures needed for flying—light bones, a furcula (wishbone), and semi-lunate (wrist) bones, but it was definitely a dinosaur. Its body was streamlined, like that of a roadrunner; its arms looked like wings with claws; and it had hairlike, thin feathers. Probably *Bambiraptor* was warm-blooded, as birds are. It preyed on small mammals and reptiles.

Until recent years, the oldest feathered fossil found was the Jurassic skeleton of *Archaeopteryx lithographica,* which is classified as a bird but had many characteristics of reptiles; like *Bambiraptor,* it is thus a link between the two groups.

Other dinosaurs varied greatly and radiated into many Mesozoic habitats. Aside from the ancestors of birds, which adapted partly by becoming smaller, the dinosaurs in general became larger. Many became fierce predators as well. The combination of large size and vicious habits served them well for a long time; during the age of dinosaurs, virtually every animal that was at least a foot long was a dinosaur.

Many of the carnivorous dinosaurs developed strong, sharp teeth and other adaptations for preying on smaller animals. (Their teeth were continually replaced as they broke off.) The ankylosaurids had tail clubs. Many

theropods had flexible lower jaws that helped them when eating large, perhaps struggling prey. Some had a powerful sicklelike digit that could be used for eviscerating prey animals. Some evolved the feathers that would be used by birds for flying, but may have originally served for staying warm, brooding young, or displaying for potential mates.

Not all dinosaurs, though, were carnivores. The herbivorous dinosaurs fed on the new plants of the Mesozoic, using beaks and teeth with good adaptations for slicing and dicing vegetable matter. Some had armored bodies that protected them from attacks by larger dinosaurs. Toward the end of the Mesozoic, the herbivores replaced the carnivores in many areas.

Plants also evolved during the Mesozoic. At the beginning of the era they were still the **seed-fern** plants that had dominated coal-age forests in the Paleozoic. By the end of the Mesozoic era both conifers and flowering plants had appeared and had begun **radiating** into many habitats. As the flowering plants evolved and spread, insects were drawn to their nectar, and the insects became adapted to compete with each other in getting nectar. As they gathered it, they inadvertently collected pollen on their bodies and spread it to other plants, which were fertilized by it. Plants that benefited from that pollination were more likely to reproduce than other plants were, so flowers that attracted insects were useful adaptations. (Today some flowers even have adapted to resemble certain female insects, a useful characteristic for attracting males of that insect species.) Thus much coevolution of insects and flowering plants occurred in the Mesozoic.

As the Mesozoic era was drawing to a close, the placental and marsupial mammals appeared on the scene, just in time to benefit from the coming extinction event that eliminated their competitors.

Figure 2.10
The Cenozoic Era.

There have been many theories—volcanic eruptions, climatic change, movement of the continents—about what events ended the Mesozoic era. One of the most likely causes, supported by much evidence, is the crash of a huge meteorite to Earth in what is now Mexico. The crash would have raised an enormous cloud of soil and rock, blocking out sunlight and cooling the climate. Although that event was sudden, the extinctions themselves took place over millions of years. (Presumably it took that long for whole ecosystems to be modified.) Eventually, however, most of the dinosaurs were gone, and the mammals moved to the front of the stage.

Cenozoic

In what is today central Los Angeles, not far from Hollywood, during the Cenozoic era tar and oil seeped to the surface of the ground. They formed sticky pools that could trap unwary animals roaming over the sunny, rolling plains. As the trapped beasts struggled, they attracted predators that also became trapped, drawing still more animals into the tar. There the animals died and conveniently became preserved for study by modern paleontologists. Today the La Brea collection of fossils is a unique record of part of the Cenozoic era.

Many of the La Brea creatures would seem familiar to you, though perhaps you would wonder about some of their characteristics, and you certainly would not see them in southern California today. The extinct western horse *(Equus occidentals)* was similar to modern horses. The bison *(Bison antiquus)* and diminutive antelope *(Breameryx minor),* too, looked somewhat unlike modern forms but are recognizable. Like the camels *(Camelops hesternus)* found at La Brea, they are startling more for their location than their anatomy.

Surprising as the idea of camels in North America may seem today, in the early

Age of Mammals they ranged widely over this continent. Today, however, their relatives are the **Bactrian** camels of Asia, the dromedary camels of Africa, and the llamas of South America; no camels remain in North America.

Other fossil **mammals** at La Brea include mastodons *(Mammus americanus)* and mammoths (*Mammuthus* species). Today their elephant descendants are found only in Asia and Africa. Fossils of large ground sloths *(Paramylodon harlani),* which probably grazed on the plains in the area, have also been found here. They were similar to the tree sloths of today's Central and South America.

In addition to all these **herbivorous** animals, there were their huge predators— saber-tooth cats *(Smilodon californicus).* These animals were about as large as modern African lions and had enormous daggerlike canine teeth, so they were no doubt ferocious beasts. Nor were they the lone carnivores at La Brea: A huge lionlike cat, *Panthera atrox,* also preyed on the elephants and ground sloths. Three kinds of bears, including *Ursus optimus* (a black bear related to living species), were present. There also were small carnivores—skunks, weasels, and badgers—closely related to those of today. Dire wolves *(Canis dirus)* probably hunted in packs, as modern wolves do.

Birds as well as mammals became trapped in the sticky tar pits and died. The huge condorlike vulture *(Teratornis merriami),* an ancestor of the living California condor, is an especially remarkable fossil. Other birds at the site included eagles, storks, grebes, and falconlike and gooselike birds. Reptiles and amphibians, some extinct and some closely related to modern species, were here, as were insects.

Although the animals at La Brea were mostly quite unlike modern animals in California, the plants seem to have been quite similar to those of today. Pine cones in the tar probably fell from Bishop pines *(Pinus muricata).* There were also coast live oaks *(Quercus agrifolia),* cypresses, and junipers, trees found still in California.

The Cenozoic era began about 65 million years ago, after the cataclysmic events that wiped out the dinosaurs. As shown by the La Brea fossil assemblage, a Cenozoic landscape may seem similar to a modern one. However, on looking more closely, few animal species are just like those of today, though the major modern groups— such as turtles, birds, crocodilians, snakes, lizards, and mammals—are present. Also, geographic distribution of many animals has changed considerably.

During the Cenozoic, plate tectonics led to the continents becoming arranged as they are today, and earlier forms of today's plants and animals continued evolving to those of the present. Mountain ridges rose, driven upward by the movements of the plates. The supercontinent Laurasia broke up, the Atlantic Ocean expanded, and the Pacific became smaller. All of the continents except Antarctica moved northward.

At the beginning of the Cenozoic, the world climate was much warmer than it is today. About 50 million years ago, cooling of the Earth began; there have been some fluctuations, but in general cooling has continued since then. (Since about 1925 we have experienced a global warming trend; that is a very short period on the geological time scale.) During the Cenozoic, glaciation developed on the Antarctic continent (about 35 million years ago) and in the Northern Hemisphere (from 3 to 2.5 million years ago).

At various times since life began on Earth, glaciers have advanced across parts of the Northern Hemisphere, carving out many physical features and leaving the Great Lakes, other lakes, and rocky, eroded earth behind them when they retreated. The so-called Ice Ages have been concentrated in four major periods: 800 to 600 million years ago, 460 to 430 million years ago, 350 to 250 million years ago, and from 4 million years ago to the present. About 10,000 to 8,000 years ago, a major extinction may have contributed to the sudden disappearance of many Ice Age mammals, such as the mastodon. (The extinction may have resulted either from climatic change following the melting of Pleistocene glaciers or from hunting by early human hunters.)

Today there is a great diversity among flowering plants (angiosperms); this is a result of much radiation during the Cenozoic. As the climates diverged and became those of today's biomes, the flowering plants adapted to live in one ecosystem or another. For example, in colder climates the deciduous trees that could withstand winter were more likely to survive than were broadleafed trees that kept their leaves all year.

After the massive extinction of dinosaurs and other organisms that began the Cenozoic, many ecological niches were left empty. Mammals, which had been of

TOOLS: RADIOACTIVE DATING

The age of ancient rocks and the fossils within them can be determined by the method called radioactive dating. It is based on the facts that rocks contain radioactive elements and that any radioactive element breaks down to other elements at a constant, predictable rate. The amounts of the original ("parent") and "daughter" elements in the rock can be measured, and the ratio between them can be used for calculating the rock's age. The time during which half of a parent element breaks down to a daughter element is called the half-life of the element. The breakdown involves the parent's atomic nucleus only, so the rate is not affected by environmental conditions, such as temperature or pressure. That makes radioactive dating a very reliable method.

Many elements, such as uranium, occur in nature as different **isotopes.** The isotopes have their own unique half-lives, so a uranium-containing rock can be studied in terms of both isotopes, uranium-238 and uranium-235. Uranium-238 has a half-life of 4.5 billion years, during which half of the original amount decays through a series of steps to lead-206. The half-life of uranium-235 is 713 million years, producing a "daughter," lead-207.

Different radioactive elements decay at different rates; carbon-14, for example, has a half-life of 5,730 years. That means that half of a given amount of carbon-14 will break down to a daughter element, nitrogen-14, in 5,730 years. Carbon-14 is useful for measuring the ages of rocks between 100 and 40,000 years old.

Another element that may be used for dating rocks is thorium-230, which is one of the products of uranium-238's breakdown. Thorium-230 itself breaks down with a half-life of 75,000 years.

Table 2.1
Isotopes Used in Radioactive Dating

Parent element	Daughter element	Half-life of parent in years	Useful range in years
Carbon-14	Nitrogen-14	5,730	100 to 30,000
Potassium-40	Argon-40	1.3 billion	100,000 to 4.5 billion
Rubidium-87	Strontium-87	47 billion	10 million to 4.5 billion
Uranium-238	Lead-206	4.5 billion	10 million to 4.6 billion
Uranium-235	Lead-207	710 million	

minor importance during the Mesozoic, suddenly radiated and filled those niches. As with the La Brea fauna, elephants, camels, saber-tooth cats, and other large mammals spread throughout the Cenozoic plains and forests. Late in the era, during the Pleistocene epoch, early modern humans appeared.

EVOLUTION BY NATURAL SELECTION

As Pangaea broke up in the Triassic and Jurassic periods, the earliest dinosaurs were separated from one another. They continued to evolve in different areas of Laurasia and Gondwana, with some new species appearing as a result of selection. For instance, fossils of familiar dinosaurs such as *Tyrannosaurus rex* and the hadrosaurs have been found only in western North America and eastern Asia. In Africa, some dinosaurs evolved to become tiny "duck crocs," *Sarcosuchus.*[9]

A group of animals or plants that are structurally similar and can interbreed is a species. Evolutionist Ernst Mayr provided this definition of species: "Species are groups of actually or potentially interbreeding natural populations, which are reproductively isolated from other such groups." Some biologists add the condition that the offspring of any crosses be fertile.

Members of a species may live in the same area, as members of a local population, or they may be widely separated. As long as they can still interbreed, they are still members of the same species. Humans all over the world, for instance, belong to the species *Homo sapiens,* though they may appear quite different from each other. The "classic Neanderthals"—those first described—are classified as *H. neanderthalensis,* a separate species. Some closely related fossils, though, are given the name *H. sapiens,* with the subspecies *neanderthalensis.* Members of different subspecies may interbreed to produce hybrids, and some of the *H. sapiens neanderthalensis* group may have interbred with other *H. sapiens.*

As a result of natural selection by the environment, some individuals survive until they can reproduce, and others perish. Selection acts on the variety of organisms on Earth at any one time. But where does that variety come from? Darwin was troubled by this; he knew that organisms must change their characteristics but not how it happened. Another biologist, Austrian monk Gregor Mendel, had found out some of the answers already, but Darwin didn't know about Mendel's work when he wrote the *Origin.*

Today, thanks to Mendel and to his twentieth-century successors, we know that organisms change by mutations (changes) in their DNA. Mutations cause differences in the proteins produced by cells, and that leads to changed characteristics on which the environment can act. Most mutations are harmful to the organism, but a small proportion survive and may give the organism an advantage in the struggle for survival. These changes that are the microscopic basis of evolution are called microevolution, in contrast to the macroevolution seen in whole organisms and populations. (Mutations are discussed in Chapter 3.)

As you have seen, species evolve over time. Their evolution results from interactions of four phenomena:

1. Every species has a potential to increase in number by the production of offspring.

2. The offspring are genetically variable, so they have variable abilities to use resources.

3. The resources needed for survival in each environment are finite.

4. Each environment selects offspring that are best able to survive and leave offspring of their own.

Variation

The millions of different species of plants, animals, and microorganisms that live on Earth today are all related to some degree. Closely related organisms have a common ancestor; that ancestor, along with others, had a common ancestor; and so on back to the earliest ancestors.

Acacia trees grow tall in Africa, some of their leaves beyond the reach of most browsing animals. Long-necked giraffes, however, can easily reach the leaves. How did giraffes come to have those useful long necks, which seem like an awkward hindrance in other ways?

French biologist Jean-Baptiste Lamarck proposed in 1809 that when an organism is challenged by its environment, it may not only adapt to meet the challenge but also pass its adaptation on to its offspring. Giraffes, for instance, are likely to stretch their necks considerably to reach the highest acacia leaves; Lamarck thought that a giraffe whose neck had been lengthened in this way might undergo a heritable change and have young with longer necks than usual. His idea, called Lamarckism or the inheritance of acquired characteristics, is appealing because it seems to reward effort, and over many generations there does seem to be evidence for it. Before Darwin, and even afterward, many biologists accepted Lamarckism.

Darwin's theory was different in that he thought natural selection operated on nondirectional variation. That is, in his view a giraffe that had stretched its neck in response to tall acacia trees might produce offspring with longer necks. But—and this is an important distinction—it might also produce offspring with shorter necks, or with necks only as long as the parent's neck was before it was stretched. The acacia-filled environment then acted on the offspring, perhaps eliminating those with shorter necks, or making them malnourished. They were unlikely to reproduce. The young with longer necks, however, would survive and leave offspring of their own. Genetics is a complex phenomenon, but a giraffe with a naturally long neck would be somewhat more likely to have long-neck offspring than would a giraffe with a naturally short neck. Thus, over many years selection could lead to change in an inherited characteristic.

Notice that the outcome is the same whether explained by Lamarckism or Darwinism. The difference is that in Lamarckism, there is a deliberate response to the environment, but in Darwinism, genetic variation occurs in many directions, and the environment then selects the most useful variations for survival.

Even Darwin speculated that organisms' hereditary material might be altered by a process he called pangenesis. He hypothesized that an organism's body cells might accumulate small particles during growth and throughout the life span. The particles could then pass from body cells to germ cells (sperm and ova), where they could pass some changes on to the organism's offspring.

The idea of pangenesis was discredited later, and Darwin deemphasized it in his writing. For him, the force of natural selection was what brought about evolution; pangenesis was just one way in which variations might appear and be inherited. Because his theory did not depend on pangenesis, it was not harmed when pangenesis lost out.

Speciation

Natural events may bring about a separation of two parts of a population. A barrier

APPLICATION TO EVERYDAY LIFE FEATURE: ANTIBIOTIC-RESISTANT BACTERIA

Half a century ago, penicillin was a fairly new antibiotic. Hailed as a **panacea** for most bacterial **infections,** penicillin has changed our way of life in the United States. People have lost their fear of many diseases like pneumonia, rheumatic fever, and syphilis, knowing that even if they are infected, penicillin can probably cure them.

Nearly 100% of bacterial infections caused by *Staphylococcus* were susceptible to penicillin in 1952. By 1982, less than 10% were susceptible. This happened because penicillin has acted as a selective force in the environment, killing susceptible strains of bacteria and allowing resistant strains to survive and reproduce.

Many susceptible strains are not only harmless but form part of the normal bacterial flora on our skin that help maintain the body's health and balance. When the normal flora are killed, the resistant strains have no competitors and can multiply.

Even when used for infections caused by susceptible bacteria, antibiotics can get into the environment. All too often, they are used **prophylactically,** for a bacterial infection that may or may not respond to them, or even for a viral infection.

When a California teenager died in 2001 after contracting meningitis, the county health department set up a special clinic where they passed out ciprofloxacin, a powerful antibiotic, to anyone in the community who felt they might have been exposed to the disease. Nearly 1,600 doses were given, and the local school district was advised not to let students return to school without proof they had taken it (Bell 2001). However, because the disease is passed among people by their sharing saliva— for example, in kissing, sharing a cigarette, or drinking from the same glass—it seems unlikely that very many of the 1,600 people were actually in danger. By taking the prophylactic drug, they may have simply killed bacteria that were susceptible to it and allowed the emergence of resistant strains.

Similarly, when some high-profile cases of **anthrax** appeared in the general panic of late 2001, not only was ciprofloxacin used rather than penicillin and other antibiotics, but people began hoarding it just in case they became exposed to anthrax spores at some time.

Today many products contain antibiotics that are not needed but that increase the levels of antibiotics in the environment. When we insist that doctors give penicillin or other antibiotics for a common cold, there may be a placebo effect that makes us feel better, but the virus causing the cold is unaffected and more resistant bacterial strains may emerge.

Source: Bell, E. 2001. "Back to School in Livermore." *San Francisco Chronicle,* 21 April.

---

may be created by an earthquake, drought, or flood. In many cases, the barrier is temporary, and eventually the two parts of the population are reunited. Sometimes, though, the barrier lasts for many years, with continued isolation of the two groups.

If the environments on the two sides of the barrier are unlike, they will select different characteristics for survival. (The different geographies and climates of Laurasia and Gondwana in the Mesozoic, for exam-ple, probably selected for different kinds of dinosaurs.) Finally, the two groups' traits may become so unlike each other that they can no longer interbreed, and they may look or act unlike. At that point, one or both populations have become different species.

ADAPTATIONS AND SURVIVAL

When an organism is suited to live and reproduce in a particular environment, it is

Figure 2.11

The upper graph shows that many individuals in a population can tolerate various levels of an environmental factor, such as heat or moisture. The lower graph shows the effects of selection for high and low levels; two populations with widely separated tolerances may eventually be formed.

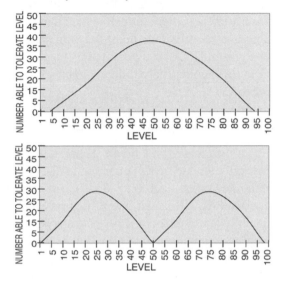

said to be adapted to that environment. For instance, in the dark abyssal regions of the ocean there are blind fish, and there are also bacteria that use sulfur instead of sunlight as a source of energy for food-making. Neither the fish nor the bacteria are adapted for life in sunlight; nearer the water surface, they would probably die. Here in the sunless, cold depths of the ocean, however, they can succeed and pass on their characteristics to the next generation.

The broad leaves of trees like oaks and maples are fine adaptations for carrying out photosynthesis in the deciduous forest, but the trees in drier areas need narrow leaves with thick **cuticles** that retard moisture loss, to compensate for the loss of moisture by evaporation. So in the mountains, the deserts, or taiga, trees like pine and spruce are adapted to withstand dryness. They photosynthesize less than deciduous trees do, but the leaves are retained for more of the year.

Figure 2.12

Speciation can follow if part of a population crosses a barrier and adapts to a new area having different environmental conditions. In (1) the population lives in Zone A, which is cold and wet. In (2) part of the population manages to cross a mountain range into Zone B, which is warm and dry. There, some individuals survive and adapt to new conditions. In (3) the individuals reproduce and spread throughout Zone B. Over time, mutations occur and give Zone B individuals other characteristics. In (4) when some of them manage to move back into Zone A, they can no longer interbreed with the population there; they have become a new species.

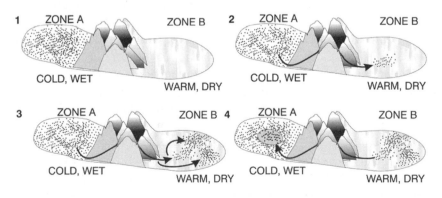

Leaves first became abundant in the plant world during the late Devonian period; before that, most plants were like leafless stems. Biologists have speculated for years about how the transition took place. In 2001, some English researchers came up with a new explanation for why plants evolved to become leafy—they were adapting to an abrupt drop in the amount of atmospheric carbon dioxide that occurred at that time. Earlier plants had few stomata, the openings through which they absorb carbon dioxide for photosynthesis and give off oxygen. Stomata also help keep the plant cool. With a high level of carbon dioxide in the atmosphere, stomata were of little importance, but when the level dropped, plants with few stomata were in danger of overheating. Selection by an environment having less atmospheric carbon dioxide would eventually have favored the development of leaves, broad structures having many stomata.

In competition with other organisms for resources, similar populations may win or lose the race because of very small adaptations. For example, Darwin noticed that both crimson clover and common red clover have tubes from which bees can suck nectar and that the tubes appear to be the same length at first glance. However, the tubes of red clover are just slightly longer. For that reason, honeybees can get nectar from crimson clover, but only the larger bumblebees can use the red clover. The difference allows both kinds of clover to be pollinated; they have different niches in the same habitat.

Useful adaptations may include behaviors or processes that affect other organisms in the community. The *Nicotiana attenuata* plants growing in Utah, for instance, are sometimes attacked by *Manduca* caterpillars. The plants respond by emitting certain **volatile** organic compounds that not only increase attacks by the caterpillars' predators, but also reduce the *Manduca* moths' egg-laying rate.

As organisms move into new areas, they may already have suitable adaptations enabling them to thrive there. Or, if new mutations arise that give them a competitive advantage, they will do well in the new environment. Often, though, new mutations are more harmful than helpful. For instance, changes in DNA brought about by radiation or toxic chemicals are unlikely to be advantageous.

Adaptive Radiation

As discussed in Chapter 1, populations naturally tend to grow larger and to disperse into new areas. As they do so, the new ecosystems will select them for survival or death. If some organisms in a species acquire adaptations that allow them to survive in a new area, they become changed. If other organisms in that species manage to invade other ecosystems, they will become changed in other ways. Eventually different populations of a species may change in many different ways as they disperse, until many new species appear. This process of change during dispersal into new environments is called adaptive radiation.

Adaptive Convergence

Conversely, the many different groups moving into the same area also are subject to the forces of natural selection; as a result, different populations—even different classes of organisms—may acquire similar adaptations that are needed in the new environment. Marine mammals, such as whales and dolphins (whose ancestors lived on land), for example, have evolved to have fishlike shapes that make them able to live in water. Swimming and diving birds, too, have streamlined shapes that aid their ability to swim and feed in the water. This process of different kinds of organisms becoming more alike as a result of selection by the same environment is called adaptive convergence.

Extinction

Whether organisms are becoming less like others of their species or more like those of other species, they are changing in response to new environmental challenges. Only if they remain for a very long time in an unchanging environment—an unusual situation—can they fail to evolve without suffering consequences.

If plants and animals cannot adapt to new conditions, they eventually die out. The death of an entire species is called extinction. Once extinct, a species can never reappear; that would require a repetition of the hundreds of millions of years of evolution that originally produced the species.

Gradual or Sudden Changes?

A major problem for paleontologists has been the gaps in the fossil record. You can search through the layers of rock in the Grand Canyon or elsewhere and find many fossils that bear mute witness to change through time, but the differences between fossils along the same evolutionary line are fairly great, even when the rock stratum seems undisturbed. Why do scientists not discover orderly sequences from one species to another in such rocks?

Darwin's View

Darwin, too, was troubled by the gaps between fossils in the same line of descent, even in layers of rock where they would be expected to appear. His picture of evolution was of very gradual change, probably at much the same rate, since life began. In the nineteenth century paleontologists were dis-

covering rich deposits of fossils from the Cambrian period, and scarcely anything at deeper layers. Darwin thought that with gradual evolution, there should have been many pre-Cambrian finds as well, but they were rare. He never completely resolved the problem to his own satisfaction, but he reasoned that the gaps must represent periods when fossils were moved to other areas (by floods or other forces), broken apart, or lost in some other way. In his view, the original transitional forms existed at one time but did not survive the geographic and climatic changes that occurred over hundreds of millions of years after the organisms died.

Until fairly recently, most biologists accepted the Darwinian position without reservations. Having a good overall record of major forms' changes through time, they accepted the lack of intervening forms as a matter of irretrievable information or as a record yet to be explored.

Punctuated Equilibria

In 1972, young biologists Niles Eldredge and Stephen Jay Gould startled the world of paleontology with a modification of Darwin's theory.[10] They suggested that mutations and visible changes in organisms proceed at a slow rate for long periods, then speed up at intervals. Gould and Eldredge referred to the process as punctuated equilibrium (now often called "punk eek"). Long periods of equilibrium would be interrupted by punctuation, just as a long paragraph may be interrupted by periods and commas. That would account for the lack of transitional forms in the fossil record—the transition would have occurred suddenly on the geological time scale.

Evidence of Evolution

The results of evolution are all around us. We can see them in the anatomy, physiology, and behavior of humans and other organisms.

Opposable thumbs are a strikingly human anatomical characteristic. (To understand just how much you depend on yours, try taping your thumbs to your palms for a few hours.) They helped make it possible for early humans to make and use tools, one of the major advances in our evolution.

Surprisingly, giant pandas also appear to have thumbs, which they use for stripping bamboo (their food). How could these bears have evolved such human-seeming appendages? Bears in general have paws ending in five clawed digits rather than the hands with four fingers and a thumb that humans have. Giant pandas, though, have five fingers plus a "thumb."

Gould wrote that the thumb of a panda is actually a padded spur, an outgrowth of the sesamoid bone. (The sesamoid is comparable to a bone below the fingers in a human hand.)

A recent study at the National Science Museum in Tokyo used MRI (magnetic resonance imaging) and other imaging techniques to shed some light to how the panda's thumb works.[11] It can't move by itself, as a human's can; but when the other five bones are curled into a grasping position, the thumb bends into place so that the panda can handle food and other objects quite well. Like many characteristics in various species, the panda's thumb is a makeshift construction from structures having other functions, serving a new use.

One of our important physiological adaptations is the fight-or-flight response to danger. If humans perceive a threat, automatically our heart rate increases, our eyes dilate, we breathe faster, and we make other responses that are useful in emergency situations. In relaxed conditions, those responses decrease and we make physiological responses for eating and using food.

Organisms have behavioral responses to internal changes and to external stimuli. Responses to external stimuli can result from interactions with the organism's own species and others, as well as environmental changes; these responses either can be **innate** or learned. The broad patterns of behavior exhibited by animals have evolved to ensure reproductive success. Animals often live in unpredictable environments, so their behavior must be flexible enough to deal with uncertainty and change. Plants also respond to stimuli.

Like other aspects of an organism's biology, behaviors have evolved through natural selection. Behaviors often have an adaptive logic when viewed in terms of evolutionary principles.

Many mothers lament the sacrifices they have made for their children, but few go so far as the mite (a relative of spiders) *Adactylidium*. An impregnated female *Adactylidium* attaches herself to the egg of a thrips, a sucking insect. The mite's eggs hatch within her body, and the larvae grow to adulthood within her, even copulating there. All their nutrition comes from devouring their mother from the inside. Eventually they cut holes in what is left of her body and emerge into the outside world.[12]

If you look at *Adactylidium* from a Darwinian point of view, you may wonder what adaptive advantage this situation has for the mother. Certainly she gets nothing out of it except for some ungrateful offspring. But that is the point. By providing nourishment for offspring she ensures that some of her genes survive and are passed on to future generations. Her behavior may seem altruistic, but it is actually selfish in that the more she sacrifices of herself, the more of her genes are likely to survive. Humans also seem to have evolved in ways that prefer the survival of our genes to our own individual survival.

The "selfish gene" idea was originated by biologists like Richard Dawkins (1941–), who wrote *The Selfish Gene,* and William D. Hamilton (1936–2000). Hamilton found that "kin selection," not individual selection, explained many mysterious animal behaviors. Kin selection refers to selection for behaviors that lead to perpetuation of an organism's genes but not necessarily to perpetuating the organism itself. It may explain all sorts of altruistic actions, such as the following.

- "Soldier" termites and ants place themselves in positions of the greatest danger when the nest is attacked. They may die while saving their relatives.
- When chacma baboon troops forage, the dominant males keep watch for predators or rivals and fight them if necessary to protect the rest of the troop.
- Many birds feign injuries and lie in front of intruders, distracting them from the birds' nests.[13]

Kin selection may include providing offspring with the best possible assortment of genes by choosing mates with good genes, or at least by choosing mates who are not relatives. (Inbreeding leads to the accumulation of deleterious genes that may cause defects or predispose the individual to disease.) Biologists have long thought that animals may choose unrelated mates by their smell—behavior inelegantly called the armpit effect. Recently animal behaviorist Jill Mateo and psychologist Robert Johnston demonstrated that the armpit effect does exist, at least in golden hamsters. By giving a group of strange hamsters a long sniff, golden hamsters can recognize their own relatives within the group.[14]

Studying behavioral biology of humans has special importance, as it provides links to much that is important in life. Scientists in psychology, sociology, education, and anthropology are often concerned with finding answers to questions about human behavior. When the selfish gene idea has been applied to humans by some biologists, it has proved very controversial. To many social scientists, as well as many laypersons, the suggestion that some human behavior has genetic roots is anathema. It seems to give people an excuse to justify sexism, prejudice against ethnic groups, and even warfare. **Sociobiologists** counter that genetic roots of much human behavior do exist and that it is better to recognize that fact and make conscious efforts to overcome some behaviors by education and legislation than to pretend they have no innate basis.

Some current evolutionary theory owes much to game theory. Evolutionary biologists often find themselves explaining animals' behaviors for surviving, winning competitions, or passing on their genes in terms of strategies that will work in different environments or in response to different opponents.

Although altruism is often an important strategy, kin selection does not always lead to a simple, knee-jerk altruistic response. There are times, for instance, when protecting one's offspring can harm one's own potential for reproduction, and it is a sounder decision to abandon the offspring than to protect them. For example, if birds see or hear predatory hawks or jays, they may stop returning to the nest to feed their chicks. (Though they might keep the chicks alive longer, if the parents are killed, they can lay no more eggs.)

A recent experiment gave new support to this idea. Cameron Ghalambor, now at the University of California, Riverside, and Thomas Martin at the U.S. Geological Survey in Missoula, Montana, studied hundreds of birds in the Northern and Southern

Hemispheres. Northern birds may survive for only one season, but those in the tropics and Southern Hemisphere may live longer; in compensation, Ghalambor and Martin found, birds living in the North tend to lay more eggs than Southern birds, because they may have only one chance. (Biologists refer to such behavior as investing more in reproduction.)

That behavior alone is interesting, but the biologists went farther, hypothesizing that the Northern birds might be more willing to return to the nest in the presence of a predator because they had already made a large investment in their chicks. They studied comparable Northern and Southern species of flycatchers, thrushes, wrens, sparrows, and warblers, and tested the parents' responses to recordings of calls from a hawk (which attacks adults), a jay (which attacks chicks), and a stuffed tanager (which is no danger at all and made a good control).

The tested birds avoided going to the nests when they heard either the hawk or jay, but there were differences in the Northern and Southern birds' responses. The jays pose more danger to chicks, and the hawks more to the adults, so the Northern birds might be expected to be more concerned about the danger from jays to their few chicks, and that was borne out by the study. The Northern parent birds were more likely to reduce feeding when they heard a jay than Southern parents were. In contrast, Southern birds were more concerned about hawks and quickly abandoned the nests to keep from being attacked themselves.

The study showed that some behavioral traits probably evolved to compensate for other traits that might lead to fewer total genes being passed on. Both the number of eggs laid and the parents' risk-taking behavior have been selected for in different environments, with the end result that the next generation's population size was maximized.[15]

The Gaia Hypothesis

Most biologists have a neo-Darwinian view of evolution (that is, one that combines Darwin's and Mendel's theories), in which organisms are rather passively selected by the environment. British chemist James Lovelock proposed a different sort of process, named for Earth goddess Gaia. According to the Gaia hypothesis, first published by Lovelock in 1979, Earth's high level of oxygen and low level of carbon dioxide were produced by early organisms. During the 1970s and later, Lynn Margulis, a biologist at the University of Massachusetts, provided microbiological evidence that supported Lovelock's hypothesis. Margulis thinks the most important events in evolution have been the mergers of organisms to form new species (see Chapter 5). In addition, Lovelock and Margulis think that the entire biosphere (Gaia) is the equivalent of an organism, with the levels of gases, temperatures, oceanic salt concentrations, and other abiotic factors being continually modified by the living things within.

THE HUMAN SPECIES

There are few measurable differences between humans and closely related animals, but obviously humans are different. We can make and invent elaborate tools, use and create language and mathematics, and use and communicate abstract ideas. When and how did these abilities arise?

The path to modern humans apparently was not a straight line, but has been more like a labyrinth, with many blind alleys and branches. However, we can get a general idea of how humans have probably evolved by looking back through groups who were closely related to us (genus *Homo*), less closely related (**hominids**), and distantly related (**hominoids**), to what might be called only shirttail relations (**primates**). Those in the genus *Homo* are generally taller, have larger brains, and appear more like us than earlier groups do. The primates include hominids and hominoids, but also include creatures that appear little like humans.

Primates

The order Primates today includes prosimians—animals like lemurs, tarsiers, lorises, and bushbabies—and anthropoids. The anthropoids have larger brains and flatter faces than prosimians and more human characteristics than any other animals. Along with humans, the anthropoids include the monkeys and great apes. The earliest primates appear in the fossil record from the late Cretaceous.

Primates have grasping hands that end in flattened nails, rather than claws. In general, compared with more **primitive** animals, they have larger eyes and rely less on the sense of smell.

Hominoidea

During the Miocene, the primate ancestors of the monkeys, great apes, and humans diverged. The great apes and humans of today are in the superfamily Hominoidea, indicating that they are closely related, whereas modern monkeys are in another superfamily that includes both Old World and New World monkeys.

Hominidae

Until recently biologists thought that about 5.5 million years ago to 6.5 million years ago African forests were disappearing, and thus some hominoids were forced to come down and live on the savanna (grasslands). Instead of swinging through the trees with their arms, they would have to run along the ground. The theory went that these new ground-dwellers were destined to become hominids (prehumans and humans); they were diverging from the hominoids that would give rise to the pongids (chimpanzees and other apes).

The theory seemed logical: Hominoids who had to run along the ground could do better if they adapted by walking upright. Not having to use their hands for grasping tree limbs, they could use them for other purposes, which might have encouraged tool use.

However, in 1994 some fossils found in Ethiopia made it seem that the theory might be inaccurate. They consisted of bones and teeth of early hominids and were about 5.8 million years old. Named *Ardipithecus ramidus kadabba,* they were about the size of modern chimpanzees.

Their discoverer, Yohannes Haile-Selassie, who was then a graduate student in paleontology at the University of California at Berkeley, said a toe bone indicated that they walked upright, and the teeth were hominid. It seems likely that they were close to the common ancestor of humans and chimpanzees.

The surprising thing about these fossils is that other evidence showed they lived in a cool, wet forest environment, not out on the savanna. If other fossil finds corroborate this one, the theory about upright walking and other hominid development occurring on savannas must be revised.

Dr. Raymond Dart (1893–1988), an anatomy professor in South Africa in the early twentieth century, was first to describe the late Miocene prehuman *Australopithecus* on the basis of a baby fossil. (It is often called the Taung baby for that reason.) Dart also was responsible for characterizing *Australopithecus* as a vicious little beast:

Man's predecessors differed from living apes in being confirmed killers; carnivorous creatures, that seized living quarries by violence, battered them to death, tore apart their broken bodies, dismembered them limb from limb, slaking their ravenous thirst with the hot blood of victims and greedily devouring livid writhing flesh.[16]

Australopithecus, still the oldest hominid fossil, has bccn found in many parts of Africa. Australopithecines walked upright and had skulls similar to those of chimpanzees. Their brains are a little larger (430 cc to 550 cc; a chimpanzee's brain volume is 340 cc to 450 cc). The australopithecines had relatively longer arms than modern humans; this may have been either an inherited characteristic from a previous tree-dwelling life or an adaptation to arboreal life. No tools have been found with the fossils. Their teeth—the molars are large and thickly enamel-coated, the canines are small—seemed like adaptations for eating plants rather than animals, giving the lie to Dart's description of them as being bloodthirsty. At least four species were present: *A. africanus, A. afarensis, A. robustus,* and *A. boisei.*

A. africanus was the species Dart discovered. It is also called a gracile australopithecine because it has a slighter build than the other species. *A. africanus* lived 2.5–3 million years ago.

A. afarensis was discovered in 1974 in eastern Africa by Richard Johanson, then at the University of California at Berkeley, and his colleagues. The young researchers named the 3.5-million-year-old fossil Lucy, after the Beatles' song "Lucy in the Sky with Diamonds." Only a little more than a meter in height, Lucy was even smaller than *A. africanus.* At this time, *A. afarensis* is the oldest known hominid species.

A. robustus, a South African fossil, was larger than *A. africanus* but smaller than *H.*

sapiens. A similar species, *A. boisei,* was found by anthropologist Mary Leakey (1913–1996), who named it Zinj (derived from an Arabic word meaning East Africa). Because it had very large molar teeth, it was also called Nutcracker Man. Zinj was a robust species and a tool-maker. Both *A. robustus and A. boisei* lived about 2 million years ago.

At this time, the exact route from *Australopithecus* to *Homo* is not understood. A few years ago, the path seemed clear, but current fossil finds are again confusing the issue. Most australopithecines are considered to be a side branch in evolution, not ancestral to humans. However, some anthropologists place Lucy at the point where australopithecines branched off in one direction and the ancestors of *Homo* in another.

Genus *Homo*

The genus *Homo* somehow evolved from earlier hominids that also gave rise to the australopithecines. The earliest known hominid having the generic name *Homo* was *Homo habilis,* a short, small-brained creature that walked upright and lived in Africa between 2 and 2.5 million years ago, while the australopithecines were still in existence.

Famed anthropologists Louis Leakey (1903–1972) and Mary Leakey found a series of African fossils they placed in the genus *Homo.* One they named *Homo habilis* ("handy man") because it was a tool-user. *H. habilis* had a brain volume of 642 cc and slightly more humanlike teeth than those of *Australopithecus.* Some workers have argued that it actually was an australopithecine, but it is usually accepted as a very early *Homo.*

The next to appear in the fossil record was *H. erectus,* which had a larger brain (about 940 cc) than *H. habilis;* fossils of this species have been found in many parts of the world. It has been called Peking Man, Java Man, Heidelberg Man, and other names. It had apparently replaced *H. habilis* by 1.5 million years ago. *H. erectus* had definitely human characteristics, including the use of fire and stone tools.

Archaic *Homo sapiens*

In 1856, a partial skeleton was found in the Neander Valley of Germany. Clearly, the bones were not those of a modern human—they were heavier, and the skull had a more flattened shape. Later similar skeletons were found in other parts of Europe, and the group was given a new species name: *Homo neanderthalensis.* Because of the skeletons they found, biologists assumed that all Neanderthals bent over and had an almost apelike appearance. Today scientists have determined that at least one old man had **rickets,** which gave him a stooped posture. The Neanderthals were actually quite similar to modern humans, similar enough that for a few years they were included in the *H. sapiens* species with us. (Their brains were at least as large as ours are, about 1,200 cc.) Today most biologists again classify them as *H. neanderthalensis.* They are a mysterious group, the objects of much study ever since their discovery.

What happened to the Neanderthals? Their remains are all 28,000 to 40,000 years old; apparently after that period, these early humans were replaced by the more advanced (recent) Cro-Magnon group who made cave paintings and had more obviously human skeletons. The Neanderthals may have been less skilled at hunting than their newer rivals, or they may have died out for some other reason. For many years, scientists thought they might have interbred with newer groups and have become some of our ancestors.

Scientists are still debating these possibilities. According to one recent study led by anthropologist Erik Trinkaus, chemical tests of Neanderthal bones showed they lived almost exclusively on animal protein, and so they must have been accomplished hunters. That would tend to rule out the explanation that newer groups outhunted the Neanderthals. As to the possibility that they interbred with other early humans, geneticist Svante Pääbo analyzed DNA from the **mitochondria** of a Neanderthal fossil and found that it was quite different from the DNA in living humans, making the interbreeding hypothesis unlikely. Because the Neanderthals lived during periods of advancing and retreating glaciers, it may be that small populations simply were unable to withstand the severe climatic conditions.

The answers about the Neanderthals are unclear. Theories about them, and about all human ancestors, are continually being revised as new fossils are discovered and analyzed.

In 2003, new discoveries in Ethiopia by Tim White and his colleagues finally provided fossil evidence to support the "Out of Africa" theory, which had rested mainly on genetic studies of modern humans. (The theory states that all modern humans are descended from a group who lived in Africa 100,000 years ago.) White's group found two nearly complete crania (skulls without the lower jaws) of two individuals and cranial fragments from another person. They strongly resembled modern skulls and lacked some features that are seen in Neanderthal skulls.[17] Using the argon-isotope dating method, the scientists dated the crania as being 160,000 years old. That seems to indicate that modern humans had evolved in Africa long before they moved on to other parts of the world and long before Neanderthals appeared on the scene.

Figure 2.13
Human Evolution. The current picture of human origins. About 30 years ago, biologists thought there was a single lineage of human evolution, from *Australopithecus* through *Homo erectus* and ending with *Homo sapiens*. With more discoveries, however, it has become apparent that the human family tree is more "bushy" than that.

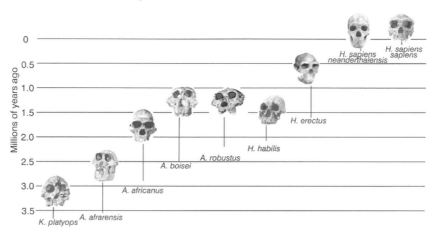

The new fossils have been named *H. sapiens idaltu.*

By the time this book is published, there will probably be newer evidence. Anthropology is a particularly exciting side of biology, continually producing new evidence and hypotheses to explain it. The history of the human species is never a "still" picture; it is always a movie.

Modern *Homo sapiens*

The *Homo sapiens* who displaced the Neanderthals were anatomically modern. Like the Neanderthals, they had large brains. Their skeletons, including the skulls, were like ours. The tools they left in eastern and southern Africa include a bola for throwing at small game, flake tools, a long flake blade, and the core from which blades were struck. We have more evidence about how they lived in Europe. The cave paintings in France especially seem to be evidence of people who used art and perhaps magic to aid them in hunting. They

made fine tools and buried their dead with some ceremony. In addition, they made clothing and were beginning to live in settlements.

Although fossil evidence is dramatic and visible, much of what we know about our human ancestors comes from DNA analysis. DNA (deoxyribonucleic acid) is the hereditary material found in the nucleus of nearly every cell that came from the person's parents. Its structure can be represented as long strings of different sequences of the letters A, G, C, and T. (The letters stand for adenine, guanine, cytosine, and thymine, chemical groups within DNA.) The more alike their DNA strings, the more closely related two individuals are.

DNA is also found in the cell structures called mitochondria. Unlike nuclear DNA, mitochondrial DNA is passed to offspring by the mother only, because the sperm cells that fertilize eggs are mainly nuclei and have no mitochondria. One result of mitochondrial DNA analysis has been construction of a family tree by geneticist Douglas C. Wallace and his colleagues. The tree begins with a human population of about 2,000 persons who lived somewhere in Africa and started to split up sometime after 140,000–150,000 years ago; this supports the Out of Africa theory.

Mitochondria are passed unchanged from mother to children, and theoretically all people should have the same string of DNA letters in their mitochondria. In practice, mitochondrial DNA has steadily accumulated changes over the centuries because of copying errors, radiation damage, and other **mutations.** Because women were steadily spreading across the globe when many of these changes occurred, some changes are found only in particular regions and continents.

Wallace discovered that almost all American Indians have mitochondria that belong to lineages he named A, B, C, and D. Europeans belong to a different set of lineages, which he designated H through K and T through X. The split between the two main branches in the European tree suggests that modern humans from Africa reached Europe 39,000 to 51,000 years ago.

Some adaptations for life in Africa, such as dark skin color, were lost by groups moving to Asia and northern Europe. There is a definite correlation between the amount of sunlight to which people are exposed and their skin color. Several theories have been advanced to account for it. One has to do with vitamin D: The vitamin is needed for calcium use and bone growth (as discussed in Chapter 4), but too much of it can lead to calcification of the arterial walls and kidney disease. Overproduction of vitamin D doesn't occur in dark-skinned people living near the equator, as the pigment in their skin absorbs sunlight, but there may have been selection against light-skinned people who produced too much vitamin D. Other explanations include protection against sunburn and skin cancer for tropical inhabitants, protection of the internal organs against overheating, protection against sudden drops in temperature in the tropics, protective coloration, and protection against frostbite by pale skin. Sexual selection for the characteristics considered beautiful in different areas of the world is another possibility. (If dark skin is considered desirable in a mate, then more dark-skinned people than light-skinned people are likely to reproduce, and the frequency of genes associated with dark skin will increase in the population.)

THE ORIGIN OF LIFE

Today biologists have reconstructed much of the history of life on Earth. That history helps explain why organisms have the characteristics they do and why they interact in certain ways with one another and with the nonliving environment. As evolutionist Theodosius Dobzhansky (1900–1975) once said, "Nothing in biology makes sense except in the light of evolution."[18] Once you begin seeing the world as the result of evolution, nearly everything does make sense. Paths from land mammals to whales can be traced, a panda's thumb seems less odd, and even much human behavior becomes more comprehensible.

Although the major patterns of macroevolution now seem clear, much remains to be discovered. Fifty years from now, perhaps, biologists will have determined exactly when humans left Africa and whether early humans also arose on other continents. They will know more about genetic relationships among evolving organisms. They may even know something about an area that is extremely mysterious now—the origin of life itself.

No one knows when humans began wondering about how life on Earth originated, but the major religions have stories about it, such as the one in Genesis. Before the time of Socrates (469–399 B.C.), the early Greek philosophers rejected mythological explanations and relied instead on rational answers to their questions about life and its origins. Thales of Miletus (c. 640 B.C.), known as the father of Greek philosophy, thought water was the basis of all things, or the "first principle." Anaximander of Miletus (c. 611–547 B.C.) thought the first principle was an undefined, unlimited substance having no qualities in itself but generating the "primary opposites," hot and cold, moist and dry. From these came everything else. The next Greek philosopher to contribute to the long debate was Anaximenes, who said the first principle was air, which thickened or thinned to become fire, wind, clouds, water, and earth.

The next, Heraclitus of Ephesus (c. 535–475 B.C.), thought the original substance was fire, and that everything begins and ends in fire. Empedocles of Agrigentum (born 492 B.C.) was first to propose the familiar idea that everything came from four elements—earth, water, air, and fire. Empedocles was also the first person to put forward the idea that species had developed by evolution.

Leucippus (fifth century B.C.) and his pupil, Democritus of Abdera (born about 460 B.C.), were the first to hypothesize that all matter is composed of atoms. Their definition of atoms is remarkably like the modern definition: small, indivisible bodies that collide and join to form objects with characteristics that depend on the number, shape, size, and arrangement of the atoms that compose them.

With Socrates, Greek philosophy became less concerned with the physical basis of life, and centuries followed before paleontology and geology made it possible to solve some of the mystery about life in the past.

Some of the first organisms on Earth appear in the fossil record of the Archaean epoch, 3.8 to 2.5 billion years ago. All of them were bacteria with special adaptations for living in the harsh conditions then. The atmosphere was probably composed of methane, ammonia, and other gases that would poison life today. During the Archaean the first photosynthetic organisms began making food and oxygen from water and carbon dioxide, changing the atmosphere enormously by adding oxygen to it. Stromatolites, colonies of photosynthetic bacteria, have been found in Archaean rocks in South Africa and Western Australia.

In 1953, chemist Harold Urey and a young biologist, Stanley Miller, worked together on what is now known as the Miller-Urey hypothesis. They reasoned that if life arose during the atmospheric conditions of the Archaean, it might be possible to reproduce the origin of amino acids (at least) in the laboratory. Preparing a mixture of the compounds water, hydrogen, ammonia, and methane, they sent an electrical current through the mixture (simulating the lightning of an Archaean storm). Amino acids were indeed produced, supporting their hypothesis. In later years, the picture of the gases in the early atmosphere changed. However, many scientists still think the first amino acids may have been generated by some similar process.

Because amino acids are the building blocks of protein, a major hurdle in forming the first life was overcome when amino acids appeared. They needed to be combined into proteins, though. Even then, it is a long way from proteins to cells. Living things have cells surrounded by membranes; they contain DNA and other chemicals. How did all the first characteristics of organisms arise?

Some scientists believe that the first life was not formed from nonliving material on

Earth, but was "seeded" here after arriving as spores or microbes from other parts of the universe. (Svante Arrhenius, 1859–1927, proposed this idea and called it panspermia.) **Meteorites,** for instance, sometimes contain compounds that could have been produced by bacteria. A 1.9-kg (4.2-pound) meteorite designated ALH184001 was found in 1984 in Antarctica by an expedition of the National Science Foundation's Antarctic Search for Meteorites program. In 1993 it was identified as having come from Mars.

A meteorite had fallen on Mars more than 4.5 billion years ago and cooled beneath the martian surface. As it cooled, the meteorite became solid rock. Later, the rock was extensively fractured by impacts of more meteorites into the martian surface. During a period when Mars was probably warmer and wetter than it is today, a fluid may have flowed into fractures in the rock and deposited carbonate minerals. Sixteen million years ago an asteroid struck Mars with so much force that pieces of the rock were flung into space. One piece wandered through space until 13,000 years ago, when it fell in Antarctica.

Scientists at NASA Johnson Space Center and at Stanford University studied ALH184001 and found evidence indicating that primitive life may have existed on Mars about 3.6 billion years ago. The team found the first organic molecules thought to be of martian origin, several mineral features characteristic of biological activity, and what might be tiny microscopic fossils of primitive, bacterialike organisms. The find has been very controversial, however, and many scientists have rejected this particular evidence for extraterrestrial life coming to Earth.

Though no extraterrestrial amino acids are yet known, some of the dust and gas around distant stars do contain complex carbon molecules (such as **benzene**) and water.

Because carbon molecules and water are important ingredients for life, their presence supports the theory that an "organic soup" of molecules from which life can arise is common in the universe. In one study, a team of scientists from Stanford University and NASA created laboratory conditions similar to those in interstellar dust clouds (by freezing and then irradiating some of the carbon-bearing molecules found there) and then analyzed the resulting chemical compounds. They found some compounds, including **quinones,** that are found in living systems and are needed for biological processes.

Even if life did not first arise on our planet, it must have done so somewhere. The search backward through time for the origin of life may be the most important scientific mission of this century.

As we have seen, evolution by natural selection occurs when some organisms can survive long enough to reproduce in an environment, and other organisms cannot. Selection acts on organisms' heritable characteristics that may be suitable adaptations to a given environment. In Chapter 3, we will examine the genetic basis of characteristics.

NOTES

1. "Marine life to recover from Galapagos oil spill," *New York Times,* January 30, 2001.

2. Loren Eiseley, *Darwin's Century* (New York: Doubleday, 1958).

3. S. J. Gould, "The Piltdown Conspiracy," *Natural History,* August 1980, 8.

4. George Bryan, *The Proverbial Sherlock Holmes*, vol. 3, no. 1, 1997, www.deproverbio. com/DPjournal/DP,3,1,97/HOLMES.html.

5. Charles Darwin, *The Voyage of the "Beagle"* (1845; London: Heron Books), pp. 379–380.

6. R. W. Lewis, "Teaching the Theories of Evolution," *American Biology Teacher* 48 (1986): 344–347.

7. National Academy of Sciences, "Evidence Supporting Biological Evolution," in *Science and Creationism: A view from the National Academy of Sciences,* 2nd ed. (Washington, D.C.: National Academy Press, 1999).

8. Loren Eiseley, "The Snout," in *The Immense Journey* (New York: Random House, 1957), pp. 49–50.

9. Damian Carrington, "Super-croc Dinosaur's Little Brother Revealed," NewScientist.com, www.newscientist.com/hottopics/dinosaurs.

10. N. Eldredge and S. J. Gould, "Punctuated Equilibria: An Alternative to Phyletic Gradualism," in *Models in Paleobiology*, ed. T. J. Schopf (San Francisco: Freeman, Cooper, 1972) pp. 82–115.

11. *Science News* 159 (January 27, 2001): 62.

12. S. J. Gould, *The Panda's Thumb* (New York: Norton, 1992).

13. E. O. Wilson, *Sociobiology* (Cambridge, Mass.: Belknap Press of Harvard University Press, 1975).

14. C. K. Yoon, "Sniffing Out the Relatives, for Survival's Sake," *New York Times,* May 2, 2000.

15. E. Pennisi, "Birds Weigh Risk before Protecting Their Young," *Science* 292 (2001): 414–15.

16. R. Dart, quoted in R. Johanson and M. Edey, *Lucy: The Beginnings of Humankind* (New York: Simon and Schuster, 1981), p. 40.

17. T. White et al., "Pleistocene *Homo sapiens* from Middle Awash, Ethiopia," *Nature* 423 (2003): 742.

18. Quoted in *American Biology Teacher* 35 (March 1973), no. 125.

3

Cells and Genetics

Realizing that natural selection had to act on hereditary differences among individuals (see Chapter 2), Charles Darwin studied many kinds of organisms in an effort to learn more about their hereditary variations. Unfortunately, he never learned about the work of an obscure Austrian monk and scientist named Gregor Mendel (1822–1884). In the 1860s, Mendel made the first systematic study of how genes behave and how they are inherited by crossing hundreds of garden peas with assorted traits and keeping track of the results. The gene was still an abstract entity to Mendel (who knew nothing of DNA), however. In fact, it hadn't yet been named. He called the units of **heredity** "factors," and early twentieth-century biologists gave them the name of genes.

Around 1900 several biologists independently rediscovered Mendel's work, and the era of modern genetics was under way. It progressed rapidly: By 1902 American biologist Walter Sutton (1877–1916) had proposed that chromosomes—large, dark-staining structures in dividing cell nuclei—were the site of Mendel's factors. The Hardy-Weinberg principle of equilibrium, showing how genes are maintained in certain proportions in populations, was formulated separately by English mathematician Godfrey H. Hardy (1877–1947) and German physician Wilhelm Weinberg (1862–1937) in 1908. In 1910 Thomas Hunt Morgan (1866–1945), an American geneticist, began his research on the fruit fly, *Drosophila melanogaster.* Morgan showed not only that genes are on chromosomes but also that if genes are very near each other on the same chromosome, they tend to be inherited together.

Since then, genetics has advanced more swiftly than any other field in biology. During the twentieth century, geneticists found that each chromosome is a long, helical molecule of deoxyribonucleic acid (DNA), and that genes are the segments of chromosomes that carry codes for proteins. They have discovered how genes **mutate** into new forms and have developed methods for altering the genetic makeup of organisms. Early in 2001, two teams of competing workers published preliminary complete maps of the human **genome,** an accomplishment of enormous importance for human health and future evolution.

THE STUDY OF CELLS

As nearly everyone knows now, individual organisms have unique sets of genes that determine all their characteristics. How many of us, though, know exactly what a gene is? How are genes related to the chemical composition of DNA and to the proteins produced by cells according to genetic specifications?

Early in the twentieth century, a series of geneticists performed observations and experiments that led to our current understanding of genes' chemical and physical structure. Investigating those aspects of genes necessarily led to new knowledge about nuclei and the cells around them.

Discovering the Subcellular World

Cells themselves had been objects of study for centuries. The first scientist to use the word *cell* was Robert Hooke (1635–1703), who was not only a brilliant biologist but also a chemist and physicist. As the seventeenth-century curator of experiments for the Royal Society in London, he performed a variety of demonstrations for the society's members.

Hooke devised a compound microscope and an illumination system for it, one of the best such microscopes of that time. With it he observed many kinds of organisms, from fleas to sponges, and he was the first person to train a microscope on fossils. In 1665 Hooke wrote a book, *Micrographia,* that contained his descriptions and elaborate illustrations of his observations.

Hooke's study of thin slices of cork led to his discovery of cells; he actually observed the empty cell walls surrounding the dead cells in cork bark. He wrote in the *Micrographia* that the bark was perforated and porous, much like a honeycomb. Hooke called the pores "cells" because of their resemblance to the cells of a monastery.

Biologists became especially fascinated with studying cells after the 1670s, when Dutch merchant Anton van Leeuwenhoek (1632–1723) built the first simple microscope. His instrument was simpler than Hooke's compound microscope, but the single lens was excellent. With it he observed a great variety of materials, such as the **protoctists** and bacteria he called "little living animalcules" in pond water. He also examined yeasts, scrapings from his own skin, urine, seminal fluid—anything he could think of that might yield information if closely inspected.

For about the next 200 years scientists were limited by the tools available for observing cells, although they made some progress. Observations by early microscopists were interesting but lacked any underlying theory. Then, in 1838, German botanist Matthias J. Schleiden (1804–1881) stated that lower plants are one-celled, and higher ones are composed of many individual cells. Later German physiologist Theodor Schwann (1810–1882) generalized the cell theory to include animals. The theory that all plants and animals are made up of cells was an important advance in biology.

A prominent nineteenth-century German physician, Rudolf Virchow (1821–1902), used the cell theory to explain the effects of disease on the organs and tissues of the body, emphasizing that diseases arose primarily in patients' individual cells. Virchow used a statement, *omnis cellula e cellula* (every cell is derived from a cell), that has become famous in biology but had actually been coined by an earlier scientist.

In the middle of the nineteenth century, chemist Louis Pasteur (1822–1895) and physician Robert Koch (1843–1910) contributed to the field now called microbiology. Pasteur demonstrated that the processes in which beer, wine, and cheeses are made depended on the actions of **microbes.** Koch **cultured** and studied bacteria, associating specific organisms with tuberculosis and other diseases.

The major force that led to modern **cytology,** though, was the development of synthetic dyes for staining cells. Natural dyes, such as indigo and saffron, had been used for centuries for staining textiles, and early workers learned that natural dyes would also react with certain cell parts, making the stained cells far more visible when looked at through a microscope. In the mid-nineteenth century, young British chemist William Perkin (1838–1907) discovered how to synthesize dyes, thereby establishing the field of organic chemistry. The synthetic dyes were more stable than natural dyes and could be prepared with a great variety of properties.

Many of the new dyes helped cytologists greatly in studying cell parts with a microscope. Some made the nuclei stand out, for instance; others stained structures in the cytoplasm (the name cytologists gave to everything outside the nucleus). A red dye, fuchsin, was especially attracted to DNA, which made it possible to see the nuclear inclusions called chromosomes. They were named by Paul Ehrlich (1834–1915), who later found the bacterial cause of syphilis.

Cell Parts

Different cell parts, or organelles, take up dyes somewhat differently because the organelles are not alike chemically. Some dyes, for example, stain acidic structures, and others stain basic structures. An organelle's chemistry is associated with its functions in the cell—such as energy production, transport of molecules, waste disposal, synthesis of new molecules, and storage of genetic material. A cell contains specialized organelles in a concentrated mixture of thousands of different molecules having different functions.

The Plasma Membrane

The outer limit of a cell is, of course, visible if the cell can be seen at all. A cell's "skin" is a double-layered plasma membrane. Early cytologists simply saw the outer surface of the cell's cytoplasm as a rather passive structure. Over the years, biochemical studies provided indirect evidence about its structure; when the electron microscope was developed in the mid-twentieth century, it showed the microscopic details of the membrane. Now it is known to be complex and to participate actively in transporting and filtering materials through itself.

Phospholipid compounds form the bilayered plasma membrane. These compounds have chemically hydrophilic ("water-loving") "heads" and hydrophobic ("water-fearing") "tails," with the tails pointing toward the interior of the membrane. Thus the outer and inner aspects of the membrane—which face an aqueous medium on either side—are both hydrophilic, but its interior is hydrophobic.

Water is a polar molecule; that is, its hydrogen atoms have a slight positive charge and the oxygen atom has a slight negative charge. Polarity makes it possible for water molecules to form **hydrogen bonds** with each other or with other polar molecules and to interact with ions. As a result, polar molecules and ions can easily dissolve in water.

In contrast, nonpolar compounds (like the tails of phospholipid molecules) are hydrophobic. For that reason the membrane is impervious to polar molecules and ions, which can cross the membrane only if

Figure 3.1
The Plasma Membrane.

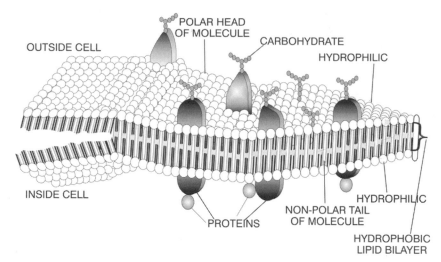

transported by **carrier-protein** molecules embedded in it. On the other hand, non-polar compounds (such as cholesterol) can easily move into and out of cells. Small molecules (such as oxygen and carbon dioxide) as well as water can cross the membrane by dissolving in the phospholipids, in a process called passive diffusion. (Compounds diffuse from an area where they are more concentrated to an area where they are less concentrated. As an example, molecules of strong perfume gradually diffuse from a perfume wearer to the other side of a room.) Facilitated diffusion is similar, but the molecules do not dissolve in the phospholipid layer of the membrane. Instead, proteins in the membrane bind the molecules and help them move across it.

Islandlike molecules of proteins and carbohydrates extend through the entire membrane. Because of this composition, the membrane can selectively absorb or act as a barrier to various chemical compounds. Thus it regulates the cell's activities and interactions with its surroundings.

Most cell functions depend on chemical reactions. Food molecules taken into cells through the membrane are broken down to provide the chemicals and energy needed to synthesize other molecules. When nutrients—proteins, fats, and carbohydrates—are carried to cells by the blood, they must pass through the plasma membrane. The membrane's bilayered structure helps make this absorption possible. The membrane also allows hormones (such as insulin), oxygen, and other important chemicals to pass into or leave the cell.

Cytoplasm

In white blood cells, bacteria may be engulfed, then taken into the cytoplasm through the plasma membrane. They are then enclosed in a membrane-enclosed sac, a vacuole. Other membranous sacs, called lysosomes, then fuse with the vacuoles containing bacteria; **hydrolytic enzymes** in the lysosomes digest the bacteria. (This process is called phagocytosis and is important in resistance to disease.)

Figure 3.2
Generalized Animal Cell.

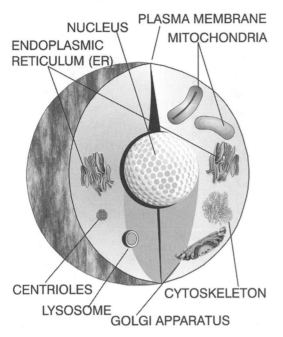

NUCLEUS PLASMA MEMBRANE
ENDOPLASMIC MITOCHONDRIA
RETICULUM (ER)

CENTRIOLES CYTOSKELETON
LYSOSOME
GOLGI APPARATUS

Vacuoles contain water and suspended materials and increase the cell's size, especially in plants. Vesicles are similar to vacuoles, but smaller. Vesicles transport materials into, out of, and throughout the cell.

Both plant and animal cells have plasma membranes, but plant cells are also surrounded by a thicker cell wall made of **pectins** and **polysaccharides.** It is stiff enough to give the plant tissue some support.

Outside the nucleus, the gel-like material of the cytoplasm is given some support by a cytoskeleton of filamentous proteins. The cytoskeleton also helps direct the movements of materials within the cell.

In electron micrographs, the cytoplasm appears to be nearly filled with something that is folded many times. This is the endoplasmic reticulum (ER), a network of sacs, channels, and tubes. Some of the ER, called

rough ER, has ribosomes—the sites of protein synthesis in the cytoplasm—at closely spaced intervals along it. The rest (smooth ER) does not. Ribosomes are the most numerous organelles in a cell.

Cup-shaped cytoplasmic structures called Golgi complexes have layers of flattened sacs, each surrounded by a membrane. Tubules and vesicles surround the complex. Lysosomes and other membranous structures are formed by Golgi bodies. Lysosomes bud off from the Golgi complex and fuse with food-containing vesicles. Other vesicles deliver lipids and proteins to the Golgi complexes, where new membrane materials are then assembled from them. Still more vesicles take the materials to the cell membrane or to other organelles, where they are used.

Numerous mitochondria are scattered throughout the cytoplasm. Like the rest of the cytoplasm, they are filled with folded membranes. Nutrients in the cytoplasm are taken into the mitochondria and broken down for energy and building materials. Their double outer membranes are folded inward in what look like shelves; the reactions of the TCA cycle take place on these surfaces. Mitochondria are most numerous in cells having high energy requirements, such as muscle cells. Mitochondria also have their own supply of DNA, which is inherited only through the mother. (A sperm cell has too little cytoplasm to contain mitochondria, so a fertilized egg's cytoplasm is all from the egg cell.)

In the mitochondria and elsewhere in the cytoplasm, breakdown and synthesis of molecules are made possible by a variety of protein **catalysts** called **enzymes.** Breakdown of some food molecules enables cells to store energy in specific chemicals that are used to carry out cell functions.

Both plant and animal cells have mitochondria, because all cells require energy;

Figure 3.3
Generalized Plant Cell.

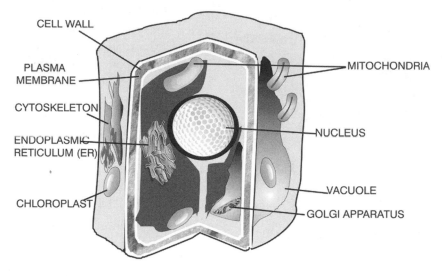

but plant cells also have other structures with inner folds, known as chloroplasts. These oval bodies contain **chlorophyll** and are the site of photosynthesis. The reactions take place on the inner folds.

Each cell in an animal contains two structures called centrioles. Formed of microtubules, they duplicate and move to the ends of the **spindles** in dividing cells. The chromosomes move toward the centrioles during cell division.

Many of the cells early cytologists studied were prokaryotes, which are very small—0.3 μm to 5 μm across—and have no membrane-enclosed nucleus. Prokaryotes include bacteria and cyanobacteria. Eukaryotes' cells are a little larger, 10–50 μm across. They have nuclei enclosed in membranes (called nuclear envelopes) and are more complex than prokaryotic cells. Prokaryotes move by lashing long, whip-like organs called **flagella,** or shorter versions called **cilia.** Cilia and flagella also are found in some animal cells.

The Cell Nucleus

Eukaryotic cells have a centrally located nucleus containing the cell's DNA. Each molecule of DNA in the nucleus coils up, shortens, and thickens to become a chromosome just before cell division. Because DNA is copied at each division, the information it contains can be passed from parents to offspring, and from the fertilized egg to all other cells in a developing organism.

All the cell's activities are directed by the nucleus. The genetic instructions coded in nuclear DNA are copied to ribonucleic acid (**RNA**) that moves out into the cytoplasm. By sending messages to the rest of the cell, DNA controls the production of thousands of proteins in the cell. The proteins then determine all the cell's functions.

Inside the nucleus is a denser spherical structure, the nucleolus. This is where the DNA is copied to **transfer RNA (tRNA),** the carrier of DNA's instructions to the cytoplasm.

Figure 3.4
Generalized Bacterial Cell.

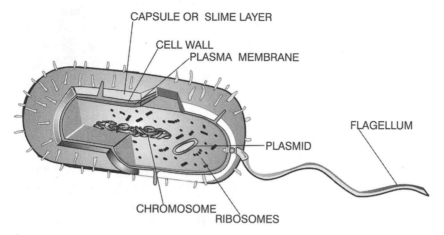

In a resting cell (one that is not undergoing cell division), the DNA molecules are spread out in a tangled mass and cannot be stained particularly well. As the cell prepares to divide, however, the molecules replicate themselves (that is, copy themselves exactly), shorten by coiling up, and can easily be stained. They become chromosomes, double-stranded structures. The two identical strands (called chromatids) are attached to each other at a point called the centromere.

Human somatic (body) cells each contain 2 copies of each of 22 different chromosomes. The members of these pairs appear identical to each other, and so they are called homologous. The homologous chromosomes, or homologs, are different at the molecular level, however. One carries genetic information from the maternal line (from the individual's mother), the other paternal (from the individual's father). Also, there is a pair of sex chromosomes that look different from each other: a body cell in a female contains two X chromosomes, and a cell in a male contains one X and one Y chromosome. The X chromosome is considerably larger than the Y. Generally an embryo's sex is determined at the time of fertilization, according to whether it is XX or XY.

Egg and sperm cells contain only 1 chromosome of each pair, or 23 chromosomes in all. That number (the haploid number, or n) is different in different species; in humans, it is 23. When an egg and a sperm unite to form a new individual, the fertilized egg again has the typical 23 pairs. Thus the diploid number for a species ($2n$, or 46 in humans) is the number of chromosomes in each body cell.

Cytologists thought for many years that human somatic cells each had 48 chromosomes. Only in 1956 was it learned that each of our cells has only 46 chromosomes. Gametes (egg or sperm cells, also called germ cells) have 23 each.

As staining techniques improved, biologists learned to prepare karyotypes (photos of the chromosomes in a single cell of a particular organism). These demonstrate clearly that the same 46 chromosomes occur in all human cells, except when there is some genetic abnormality. For example, any somatic cell from a person with Down syndrome has the usual com-

Figure 3.5
The banding patterns on two human chromosomes, 7 and X.

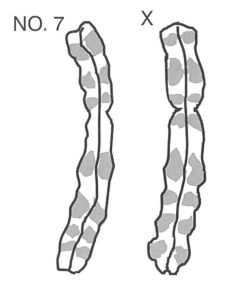

NO. 7   X

plement of chromosomes plus an extra chromosome 21.

When the chromosomes appear during the early stages of cell division, they can be distinguished from each other not only by size and shape but also by the pattern of bands on each chromosome. (The bands are not genes. A single gene is too small to be visible with a light microscope.)

The genes are segments of the chromosomes, as are long strings of noncoding or "junk" DNA. (The functions of the junk are still unknown. It may be redundant information that provides insurance against damage, it may have some other use, or it may really be junk.) Usually a gene is defined as enough information for the cell to assemble one protein from amino acids in the cytoplasm. (Sometimes it is defined in terms of providing information for one RNA segment.)

Any gene is symbolized by a series of letters that stand for the chemical units making up the DNA molecule. The letters are the initial letters of four bases—adenine (A), guanine (G), cytosine (C), and thymine (T). Each triplet of bases, such as AGT or CTG, tells the cell to add a certain amino acid to a growing chain that will eventually be a protein specified by the gene. So, a gene consists of a series of triplets.

Different forms of the same gene are called alleles. Two alleles for human eye color, for example, are the DNA codes that ultimately lead to brown eyes or blue eyes. Most genes can have several different alleles, but only one of them can be found at that gene's locus on a chromosome. Because each chromosome has one allele at that locus, the cell might have two alleles for brown eyes, two for blue, or one for each color. (Other eye-color alleles are also possible, but only two can be found in any individual person.)

Now consider an allele with an abnormal outcome, the gene for Huntington's disease (formerly called Huntington's chorea). It is found on human chromosome 4. Huntington's is an ultimately fatal disorder that begins with uncontrollable body movements and mental deterioration, caused by the death of **neurons** in the brain's **basal ganglia.** It is fortunately not common, but does afflict about 30,000 Americans. Every somatic cell has a chromosome 4 pair. The abnormal form of the gene that causes Huntington's is **dominant** over the normal form, and so if either of the pair has the abnormal allele, the person will have the disease.

The Huntington's gene doesn't produce the disease directly but brings it about indirectly by altering the normal formation of a protein. Normally, the Huntington's gene contains 11–34 repetitions of the triplet CAG. Because of a heritable mutation (change) in the gene in Huntington's victims, there are 35–100 or more of those triplets. The result is that brain cells either can't make a specified protein at all or the protein is shaped wrong and can't function properly, which leads to the cells' death.[1]

MUTATIONS

Changes in DNA (mutations) occur spontaneously at low rates. Although some of these changes make no difference to the organism, others—such as the allele that leads to Huntington's disease—can change cells and even organisms. Mutations in somatic cells can damage the cells enough to cause cancer or other disorders in the individual, but only mutations in gametes (sperm and egg cells) can create variations that are passed on to an organism's offspring.

Different chromosomes arise as mutations in normal chromosomes. A slight alteration in the DNA of a gene can produce a new chromosome that codes for a new protein and leads to an altered characteristic.

GAMETES AND SOMATIC CELLS

Though each human body cell has 46 chromosomes, or 23 pairs, each of the gametes (sperm and eggs) only has 23 chromosomes in all. Instead of the 23 pairs in a body cell, each gamete has just one chromosome of each pair. Because it is a matter of chance which of the individual's chromosomes end up in a particular gamete, the number of possible combinations of chromosomes in any human gamete (8,388,608, or 2^{23}, or 2^n, where n = the haploid number) is enormous; variation in the genes can also arise in other ways during the formation of gametes.

When a particular sperm and egg unite to form a zygote, or fertilized egg, the variations are multiplied even more. Alleles from the father, combined at random in his sperm, join alleles from the mother, which have also been combined at random from her own stock of alleles.

Biologists think this variation is the answer to the question of why so many organisms reproduce sexually. Why have many organisms, humans among them, accumulated all sorts of physical and behavioral adaptations ensuring that sexual reproduction will come about? By continually producing offspring having new combinations of alleles, sexually reproducing organisms make it possible for natural selection to choose the offspring most fitted for survival in a changing environment (see Chapter 2). Many plants and fungi may do very well using asexual reproduction, but they vary little; if the environment changes, they may not be able to survive.

Cell Division

Though a parent organism has two genes for each trait, the sperm or egg cells produced by the organism have only one. That difference results from a special form of cell division occurring during formation of the gametes (sperm or eggs), called meiosis.

Meiosis is a modification of mitosis, the process used by most cells to form new cells. Mitosis is not a simple splitting in two, but a complicated series of steps. The five major steps are called interphase, prophase, metaphase, and telophase.

Mitosis

Interphase is the condition of a cell that is not dividing; it is growing and active in other ways, however. In the nucleus, the DNA is not condensed into chromosomes but is strung out throughout the nucleus. As interphase ends and the cell nucleus prepares to enter prophase, all of the nuclear DNA is copied exactly, or replicated. This exact copying makes it possible for each of the future two daughter cells to receive the same genetic information.

APPLICATION TO EVERYDAY LIFE FEATURE:
SCREENING FOR A GENETIC DISEASE

Babies who had the disease might appear perfectly normal for a few months. Before their first birthday, though, they stopped smiling; that was the first symptom of the dreaded hereditary **Tay-Sachs disease**. From then on, blindness, seizures, and paralysis followed. The children usually died by the age of five.

As yet there is no cure for Tay-Sachs, but a simple blood test now allows potential parents to know whether they are carriers of the gene that causes it. The gene leads to a lack of the protein hexosaminidase A, otherwise known as enzyme HexA. HexA breaks down a harmful fatty substance (ganglioside GM2) in nerve cells. Without the enzyme, the ganglioside gradually accumulates and eventually destroys the entire central nervous system.

Tay-Sachs disease is inherited as a recessive trait; a person can be a heterozygous carrier and be perfectly healthy, but if two carriers have a baby, there is a 25% chance that the baby will be homozygous recessive and have the disease. (Also, there is a 25% chance that it will be homozygous dominant and a 50% chance that it will be a heterozygous carrier.) It is rare in most populations, but widespread in Ashkenazi Jews (Central and Eastern European Jews and their descendants).

Because Jews have tended to marry other Jews for faith-based reasons, the gene has remained in their gene pool. Even though many babies have died, and their genes were removed from the pool, the gene has been maintained in the many heterozygous carriers.

If people learn they are carriers, they can use that knowledge in choosing partners and in making decisions about pregnancy. Some carriers may not even wish to marry other car-

riers and risk having a baby with Tay-Sachs. Others may want to monitor pregnancies and possibly abort a fetus that is destined to die before school age.

About 30 years ago the Jewish world community became active in providing screening at schools and other places where many young adults gather. They provide free or low-cost blood tests for enzyme HexA; the results are confidential, allowing people to use the information as they wish. The results of this proactive approach to controlling the disease have been amazingly successful: The incidence of this disease has dropped more than 95%, and most doctors today have never seen a case of it (Kolata 2003).

Encouraged by this success, many have suggested using a similar method for controlling other diseases that tend to afflict Ashkenazi Jews—neurological disorders, such as familial dysautonomia, Canavan disease, Niemann-Pick disease, and mucolipidosis IV; and other diseases, such as cystic fibrosis, Gaucher disease type I, Bloom syndrome, and Fanconi anemia. Although it seems like an obviously beneficial approach, some people worry about possible negative outcomes, such as stereotyping of Jews as people afflicted with many hereditary illnesses. Also, the information may not be entirely useful; for instance, in some diseases a person inheriting the genes for the disease will not necessarily become ill. Two persons diagnosed as being carriers might decide to marry or not, and a pregnant woman might find it very difficult to decide whether to abort a fetus having the genes for the disease. Although the possible benefits seem to outweigh the drawbacks, genetic screening is not a simple issue.

Source: Kolata, G. 2003. "Using Genetic Tests, Ashkenazi Jews Vanquish a Disease." *New York Times,* 18 February.

**Figure 3.6
Mitosis.**

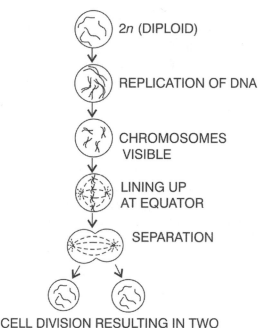

2n (DIPLOID)

REPLICATION OF DNA

CHROMOSOMES VISIBLE

LINING UP AT EQUATOR

SEPARATION

CELL DIVISION RESULTING IN TWO DIPLOID (2n) DAUGHTER CELLS

that is perpendicular to the nucleus. Microtubules are attached to the centromeres.

Anaphase begins with the centromeres dividing. They are pulled toward the centrioles by the microtubules attached to them, with one member of each chromosome pair moving toward one end of the cell, and the other member moving toward the other end. Because each chromosome had been replicated before mitosis began, each new cell receives the 2n number, like that of the parent cell.

Last, in telophase, the cytoplasm is divided between the two parts of the cell as a new nucleus forms at each end.

As the result of mitosis, each daughter cell has the normal 2n number of paired chromosomes. The two new nuclei are identical to the parent cell that produced them. Though the new cells may be smaller, they grow to the parent cell's size during the interphase that follows.

When mitosis begins, the cell enters the prophase. The DNA becomes condensed into a certain number of chromosomes; the number is the same in every body cell of an organism, with the number depending on its species. For example, in every human body cell there are 46 chromosomes, or 23 pairs. They have certain shapes, which are always the same when the chromatin (nuclear DNA and associated proteins) condenses and becomes visible under a light microscope. As a result of the replication that occurred in interphase, each chromosome consists of two identical strands, or chromatids. These are joined by a **centromere.**

At the same time, the nuclear envelope begins to disappear, and a diamond-shaped structure called a spindle is formed from the centrioles and microtubules.

In metaphase, the chromosomes line up along the "equator" of the nucleus, in a plane

Meiosis

Meiosis also begins with the replication of all the DNA in a 2n cell that is destined to produce gametes. During prophase, members of each homologous pair move close together and even twist around each other. At this time DNA from either chromosome may be exchanged with that from the other, a process called crossing over. As a result, genes from the father's side of the family may be exchanged with those from the mother's side, and the individual chromosomes are no longer maternal or paternal. As in mitosis, the nuclear envelope disappears, and a spindle forms.

During metaphase, the homologs move to the equator of the nucleus, where they take positions across from each other, and the centromeres become attached to microtubules of the spindle. Unlike what happens in mitosis, the centromeres do not divide.

Figure 3.7
Meiosis.

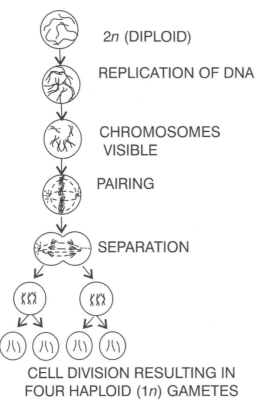

2n (DIPLOID)

REPLICATION OF DNA

CHROMOSOMES
VISIBLE

PAIRING

SEPARATION

CELL DIVISION RESULTING IN
FOUR HAPLOID (1n) GAMETES

Led by the centromeres, the homologous chromosomes are pulled toward opposite ends of the dividing cell in anaphase.

In telophase, the cells are dividing, and the homologs have completely separated. Each of the two new cells forming has the haploid number of chromosomes, but they are the replicated structures made during interphase.

Another division now begins, with the chromosomes in each new cell again moving to the equator. Because there are no longer any homologous pairs, they simply line up singly, as in mitosis. The spindle fibers (microtubules) attach to each centromere. This time the centromeres divide and lead the replicates to the ends of the cell. The final result of meiosis is four cells,

each containing the haploid number of chromosomes.

Though four sperm cells come from each meiotic process, only one egg is formed in this way. All the cytoplasm from the original cell is retained, first in one of the two cells formed in the first division, then in one of the final four. The other three nuclei are in small polar bodies that are usually cast off. This conserves cellular material to be used by the egg.

One important result of meiosis is the reduction of the diploid number to the haploid number in gametes. If this did not occur, and each gamete were 2n, the fertilized egg would be 4n rather than the normal 2n species number.

The other important event during meiosis is genetic recombination, or rearrangement of genes and chromosomes. This is the source of much variation, even within families. During prophase, genes may be exchanged between chromosomes; during anaphase, mostly maternal or mostly paternal chromosomes move at random to the new cells. When two unique gametes then unite to form a fertilized egg, a unique new individual is created.

EXPLAINING VARIATIONS

When one population in a species becomes different enough from the other populations that individuals from one can no longer interbreed with those from the other, a new species has been formed. (Different species may occasionally mate and produce offspring, but not successfully. The offspring may die or may be sterile. For example, a mule is the sterile hybrid offspring resulting when a horse and donkey mate.)

The differences among different species are clear-cut and often are evident from a casual inspection. Even within a

Table 3.1
Characteristics Used by Mendel.

Characteristic	Possible Traits	
Seed shape	round	wrinkled
Seed coat color	colored	white
Seed color	green	yellow
Pod color	green	yellow
Pod shape	inflated	constricted
Flower position on plant	axial	terminal
Stem length	long	short

species, however, populations differ in many ways; within a population, many differences among individuals can be found. The differences may be anatomical, physiological, or behavioral, but they all depend on the different proteins made by cells, and which proteins are made depend on the genetic code. Differences in the genetic code among individuals of a population are known as variations.

Mendelian Genetics

Mendel studied the inheritance of variation in the peas that grew in his monastery garden by crossing plants with conspicuously different characteristics and studying their offspring. Peas are ordinarily self-fertilizing, so he had to transfer the sperm-containing pollen from one plant's flowers to the egg-containing pistil of another's. In that way, he could be sure that he crossed, for example, a plant having round seeds with a plant having wrinkled seeds. Mendel used a shorthand in recording results for the generations; the plants he crossed were the parental, or P, generation, and their offspring were the filial, or F1, generation. When he made further crosses among F1

individuals, they gave rise to an F2 generation, and so on.

Genotypes and Phenotypes

The peas in Mendel's garden had certain **characteristics,** such as seed shape and color. He chose seven characteristics for his own research purposes; for each of these characteristics, an individual plant had either one **trait** (such as round seeds) or another (such as wrinkled seeds). Table 3.1 shows the characteristics Mendel used in his study and the pairs of traits possessed by various plants.

For several years, Mendel developed "true-breeding" plants. That is, if any two plants with the same trait were crossed, all their offspring would have the same trait. When he made crosses from true-breeding plants, he could be sure that his data wouldn't be contaminated by plants with other inherited traits. Today we call true-breeding organisms homozygous for a trait.

Although crosses of homozygous parents always produced homozygous offspring for the trait being considered, crosses between parents with different traits had more unusual results—results that did not show up in the F1 generation. For example, if homozygous green-podded plants were crossed with homozygous yellow-podded plants, all the offsprings' pods were green. But then, if some of the F1 plants were crossed with each other, they produced an F2 generation in which 75% had green pods and 25% had yellow pods. Obviously the factors determining the yellow pods had neither disappeared nor mixed with the green-pod factors but were somehow transmitted from the P1 to the F1, and thence to the F2 generation.

By counting hundreds of such crosses, Mendel found that the green:yellow ratio in the F2 generation was always the same— 75:25, or 3:1. He was the first biologist to apply statistics to studies of inheritance.

Mendel discovered that whatever inherited factor the plants possessed could be "dominant" or "recessive." In the case of pod color, the factor for green was dominant, and the factor for yellow was recessive; a plant having both factors would have green pods, because the dominant factor masked the recessive one. (He abbreviated the factors G for green, and g for yellow, to indicate that difference.) That was a useful distinction for the traits he studied, which occur in only one of two forms, but the terms can cause confusion: We know now that most characteristics are found in several forms, not just two; we call their inheritance multifactorial. Also, in some cases neither factor is dominant. The appearance of the F1 generation may be different from that of either parent in the P generation.

Mendel realized that each true-breeding plant must have two copies of the factor for each trait, and that each parent contributed just one of those copies to the offspring (which would have two copies, one from each parent). For instance, a parent that was true-breeding for round seeds *(RR)* would always provide an *R* factor, and a parent that was true-breeding for wrinkled seeds *(rr)* would always provide an *r* factor. The offspring would be *Rr*. That is, the offspring would have a genotype of *Rr;* their phenotype, or surface appearance, would be round-seeded. The hypothesis that each individual has pairs of factors for each trait and that the members of the pair separate during formation of gametes is Mendel's first law, the principle of segregation.

Some of these traits were inherited independently of other traits. For example, a plant with round seeds might have either yellow or green seeds.

Mendel realized that during formation of gametes, the factors for different traits did not stay together but behaved independently. This is now known as Mendel's principle of independent assortment. Mendel published his findings in 1866, but scientists at that time paid little attention to them. Mendel himself was well aware of their significance, saying at the beginning of his work:

Those who survey the work done in this department will arrive at the conviction that among all the numerous experiments made, not one has been carried out to such an extent and in such a way as to make it possible to determine the number of different forms under which the offspring of the hybrids appear, or to arrange these forms with certainty according to their separate generations, or definitely to ascertain their statistical relations. It requires indeed some courage to undertake a labor of such far-reaching extent; this appears, however, to be the only right way by which we can finally reach the solution of a question the importance of which cannot be overestimated in connection with the history of the evolution of organic forms. The paper now presented records the results of such a detailed experiment.[2]

Understanding mitosis and meiosis and applying them retrospectively makes it clear what happened with Mendel's factors. If each factor is on one member of a pair of chromosomes, then meiosis will produce sperm and egg cells with one factor (allele) per trait. If the trait is seed color, for instance, and the parent plant is *Gg,* then about half the sperm or eggs will contain the *G* factor, and half the *g* factor.

Chromosomes and Genes

After the rediscovery of Mendel's work, the gene did not remain abstract for long. Walter S. Sutton, a graduate student at Columbia University, was studying the formation of sperm cells in grasshoppers at the beginning of the twentieth century. He observed that as cell division began, chromosomes lined up in pairs at the cell's midline, and he reasoned that if the chromosomes were where the genes were located, Mendel's principle of segregation of factors would be accounted for. Sutton published his hypothesis, but experimental evidence was needed to support it.

Thomas Hunt Morgan provided the evidence, using fruit flies. He performed crosses of a white-eyed male *Drosophila* with a red-eyed female, finding that the F1 generation of flies were all red-eyed. When he reared an F2 generation, he had the surprising result that instead of a 3:1 red-eyed:white-eyed ratio, the ratio was closer to 4:1, and all the white-eyed flies were males. Morgan realized that the *Drosophila* alleles for eye color must be carried on the X chromosome. Because all *Drosophila* males have only one X chromosome (it is paired with the Y chromosome, which we know now carries few alleles), any allele on the X chromosome will be evident in the male phenotype without regard to dominance. In females, however, an allele on either X chromosome will behave as either a dominant or a recessive. So a red-eyed female might be either *RR* or *Rr,* but a red-eyed male must be *R*Y, and a white-eyed male must be *r*Y.

Morgan's important work (for which he received a Nobel Prize in 1933) showed not only that genes are carried in a series on each chromosome but also that some characteristics are sex-linked. Many human characteristics, too—such as male-pattern baldness and red-green colorblindness—ordinarily appear only in males because they are brought about by alleles on the X chromosome.

The picture of inheritance that emerged from Mendel's work was one of characteristics that were determined by pairs of alleles, with one member of each pair being dominant and the other recessive. Mendel had carefully chosen traits to study, though. Many traits are produced by genes with more complex alleles and actions.

In some cases of codominance, heterozygotes show the effects of both the dominant and recessive alleles. For example, human blood type is determined by three possible alleles of the blood-type gene—A, B, and O. The human blood type AB results from the alleles A and B, which are codominant. The A allele adds the sugar galactosamine to the surface of the person's blood cells; the B allele adds the sugar galactose instead. Type A people are either AA or AO; type B, BB or BO; type O, OO. Either allele A or allele B is dominant over O, but A and B are codominant.

Some heterozygous phenotypes appear intermediate between the two homozygous forms of the parents. For example, a homozygous red-flowering snapdragon *(RR)* crossed with a homozygous white-flowering snapdragon *(rr)* will give rise to an F1 generation of pink-flowering plants that look like a blend of the red and white flowers. The alleles are actually *Rr,* and that becomes apparent when the F1 plants are allowed to self-pollinate. This phenomenon is called incomplete dominance.

Mendel's pea plants had one pair of alleles determining each trait, but that is actually an unusual condition. In most cases, multiple alleles of a gene are possible, though in any individual, only two of the alleles can be present at that gene's location on the pair of chromosomes. Hair color in mice, for example, is determined by a single gene, but there are multiple alleles—albino, gray, agouti, brown, black, and others—of that gene. Furthermore, each allele may be dominant or recessive, depending on which other allele is present.

Many traits are determined not by just one pair of alleles but by the interaction of many genes—or by polygenic inheritance. Or, in pleiotropy, one gene can affect many parts of the phenotype that appear unrelated. This is because the protein produced by the gene's action may be involved in many different reactions in the body.

Genes are expressed in the phenotype as a result of interactions with their environments—which includes both the external environment and other genes. Some flowering plants, for instance, have tiny flowers and grow close to the ground in the alpine tundra. The same kinds of plants grow large in warmer, moister areas, though their genes are the same as those of the alpine plants.

Within a family, one individual may show more or less of a gene's effect than can be seen in other family members; this outcome is called variable expressivity. Some individuals in a population may not show any effects of an allele known to be present, which is called incomplete penetrance.

X-Linked Traits

Duchenne muscular dystrophy is a hereditary disease that occurs almost entirely in males; about 1 in 3,500 newborn boys have it. The gene responsible is on the X chromosome, so only in the rare event that a female inherits the allele for the disorder from both her mother and her father will she have the disorder. Traits of this type are called X-linked.

Early in embryonic development of a female mammal, one or the other X chromosomes is inactivated in each somatic cell that has been formed at that point in development. In some of the cells one X chromosome is the one inactivated; in the rest of the cells, the other X chromosome is affected. However, when any of those cells divides, the daughter cells have the same pattern of inactivation. Thus a female mammal is a mosaic of cells having one or the other X chromosome expressing itself. One striking example of X inactivation is tortoiseshell cats, which are nearly always female. The alleles for black or yellow coat color are carried on the X chromosome, and so the cat's coat is a patchwork of the two colors.

Y-Linkage

The Y chromosome has been useful for biologists tracing the dispersal of various human populations during history. This chromosome is passed unchanged from father to son. The DNA in it may be altered occasionally by a mutation, and when that occurs, all succeeding generations have the altered Y chromosome, which acts as a genetic signature for that branch of the family tree. Elaborate family trees can be constructed according to the patterns of the Y chromosome.

One group that has spread worldwide from their eastern Mediterranean origin are Jews. Genetic analysis, based on the Y chromosome, shows that Jewish populations have remained isolated from other groups even into modern times. They also closely resemble Palestinians, Syrians, and Lebanese, which suggests that all the

Semitic people descended from a common ancestral group that lived in the Middle East about 4,000 years ago.[3]

Nondisjunction

The separation of chromosomes and chromatids normally occurring during mitosis or meiosis may go awry in some cases, with the result that the daughter cells receive too many or too few chromosomes. In Down syndrome, the best-known case of this nondisjunction, a viable embryo is produced, but it has three copies of chromosome 21. Serious or mild mental retardation, increased susceptibility to infections, and various physical abnormalities characterize the victims of the syndrome. Most cases of nondisjunction are even more severe, with the fetus being spontaneously aborted during pregnancy.

In other cases, part of a chromosome may be lost during cell division; this is known as a deletion. If a deleted part of one chromosome is transferred to another chromosome and becomes part of it, that event is known as translocation. In a special type of Down syndrome called translocation Down syndrome, for example, the individual has three copies of chromosome 21, but one of them is attached to another chromosome. This produces the same abnormalities as in any case of Down syndrome.

Mitochondrial DNA

Both mitochondria and chloroplasts contain their own DNA; it is now thought that these organelles began as prokaryotes that were absorbed by larger cells and developed a **symbiotic** relationship with them that led to a permanent arrangement. Because mitochondria can be inherited only from an individual's mother, mitochondrial DNA has been useful for investigators looking at the evolution of various groups (see Chapter 2).

Jumping Genes

Barbara McClintock (1902–1992) spent much of the 1940s and 1950s studying the genetics of maize (Indian corn), looking especially at the colors of kernels and leaves. Long before the structure of DNA was established, she proposed that elements could jump around within and among chromosomes, disrupting the expression of genes they entered, changing the sequence of bases in the DNA, or enhancing the expression of other genes. She called the elements jumping genes. Though her work was ignored for many years, by the 1970s the elements—now called transposons—had been found in other organisms as well. McClintock received a Nobel Prize for her work in 1983.

John Merrick, the Victorian "Elephant Man," may have been a victim of neurofibromatosis, which is caused by a transposon, *Alu.* The transposon turns off a gene that ordinarily regulates cell growth.

A transposon called the R factor is a group of genes that produce enzymes that can break down antibiotics, such as penicillin and streptomycin. When a bacterium picks up the R factor in its genome, it is unaffected by the antibiotics. Partly because of spread of the R factor, tuberculosis and diphtheria are again causing illness in modern times.

In addition to spreading among bacteria, transposons may move among bacteria, plants, and insects. It is even suspected that a fungus that produces the anticancer drug paclitaxel acquired the necessary genes for that ability from the Pacific yew tree on which it grows.

Biochemical Studies

While Morgan, Sutton, and other geneticists were elucidating the gene–chromosome link, biochemists were doing studies that shed light on how genes operate. Today this kind of work is referred to as molecular genetics, because it focuses on the DNA molecule and its chemical actions.

Though it was clear from Morgan's studies that the genes were made of chromosomal material, questions remained. Chemical analysis of chromosomes showed they were composed of about equal amounts of DNA and protein. Because proteins are more complex than DNA, they seemed more likely than DNA to be the carriers of all the complex information known to be in genes—every physical characteristic of an organism, even all the physical bases of its behavior. Although DNA contains only four different kinds of molecular units called nucleotides, proteins are long **polymers** made from 20 different kinds of amino acids, a difference that should mean proteins can store and use much more information. Some scientists thought it probable that the chromosomes contained master models of all the enzymes and other proteins the cell might need.

A puzzling observation of bacteria provided one clue. In 1928 English bacteriologist Frederick Griffith (1877–1941) was working with *Streptococcus pneumoniae* bacteria, hoping to prepare a vaccine against the disease-producing form. That form is surrounded with a polysaccharide capsule, whereas a harmless form has no such capsules. In 1928 Griffith found that extracts from killed encapsulated streptococci could change the living, harmless bacteria to the disease-producing virulent type. They could then not only make capsules but also transmit all their new characteristics to future generations. He named the change transformation.

In 1944, O. T. Avery (1877–1955) and others at Rockefeller University showed that the extract from the virulent bacteria was DNA. Similar results were later obtained with other types of bacteria.

Avery's work identified DNA as the genetic material, but some doubts remained about whether a protein or other impurity was actually responsible for the bacterial transformations.

Chemist P. A. Levene (1869–1940) had studied DNA in the 1920s. He found that it was made of nucleotides and that each nucleotide was made of one of the nitrogenous bases adenine (A), thymine (T), cytosine (C), and guanine (G); a five-carbon sugar molecule; and a phosphate group.

Bacteriologists Max Delbrück (1906–1981) and Salvador Luria (1912–1991) worked with the viruses called bacteriophages (because they "eat" bacteria) in the early 1940s. They found that phages would infect bacteria and multiply within them until they filled the cells, which then burst open. They also found that the phages consisted entirely of DNA and protein, like chromosomes.

Continuing the work with bacteriophages early in the 1950s, American biologists Alfred Hershey (1908–1997) and Martha Chase (1928–2003) used radioactive isotopes as labels on the phages' protein and DNA. Taking advantage of the fact that protein contains sulfur but not phosphorus, and DNA contains phosphorus but not sulfur, they labeled some phage with radioactive sulfur and some with radioactive phosphorus before infecting cells. They found that the labeled phosphorus was in the infected cells and the new viruses that were produced, an indication that the viruses' genetic material was DNA rather than protein. (In later years studies with the electron microscope confirmed the Hershey-Chase experiments, showing that the phage pushes its "tail" into a cell it is infecting, injecting DNA into the cell and leaving its empty protein coat outside.)

Alfred Mirsky (1900–1974) conducted studies showing that in any species the somatic cells contain equal amounts of DNA, and the gametes contain only half as much. This is the result that would be expected from meiosis, in which the gametes have only the haploid number of chromosomes and the somatic cells have the diploid number.

Erwin Chargaff (1905–2002), an Austrian biologist working in the United States, found that when he analyzed the nitrogenous bases in DNA from several different species, adenine and thymine were present in about equal proportions and so were guanine and cytosine. Adenine and guanine are both in the class of bases called purines, which have a distinctive shape that allows them to form two hydrogen bonds. Cytosine and thymine, in contrast, are both pyrimidines, bases whose shape allows them to form three hydrogen bonds. This was a critical bit of evidence used by Watson and Crick when they built their model of DNA (discussed later).

The first successful experiments relating genes to enzymes were carried out on the bread mold *Neurospora crassa* by George W. Beadle (1903–1989) and Edward L. Tatum (1900–1975). They collected a variety of mold strains that differed from the parent strain in their nutritional requirements. Because they had undergone a mutation (change) in their genetic makeup, each needed a different particular amino acid not required for growth by the parent strain. Beadle and Tatum showed that each mutant had lost an enzyme essential for the synthesis of one amino acid.

In 1908 English physician Sir Archibald Garrod (1857–1936) set forth a new concept he called inborn errors of metabolism. These were hereditary disorders in which the body couldn't perform some normal chemical process. Later scientists realized the errors resulted from something going wrong in the production of enzymes when genes were abnormal.

Population Genetics

For visualizing results of crosses between individuals, a Punnett square is useful. The possible alleles of the sperm are shown along one side of the square and those of the egg along the other side. Each cell in the square represents a possible zygote, or fertilized egg, containing one of the four combinations of alleles.

	B alleles in sperm	*b* alleles in sperm
B alleles in eggs	*BB* fertilized eggs	*Bb* fertilized eggs
b alleles in eggs	*Bb* fertilized eggs	*bb* fertilized eggs

Punnett squares are even more useful when a whole population of organisms is being considered. Suppose that in a certain population of 100 humans, 75% have brown eyes and the other 25% have blue eyes. (The gene for eye color has several alleles, but assume that there are only these two.) The brown allele (B) is dominant, and the blue allele (b) is recessive. So a brown-eyed individual (B–) might be either BB or Bb. Assume that one-third of the B– individuals are BB, and two-thirds are Bb. Then 25 persons are BB and make only B gametes; 50 are Bb and make half B and half b gametes; and 25 are bb with only b gametes. That means that in the entire population, half the gametes produced are B, and half are b. In other words, the population gene pool is 50% B and 50% b. These are called the frequencies of the alleles.

Fitting these data into a Punnett square, you can see that it is very like a cross of one pair of parents, but now the entire population's gene pool is being crossed. In this very simple example, the resulting offspring look just like those in the P generation—25% *BB*, 50% *Bb*, 25% *bb*.

	0.5 *B* alleles in sperm	0.5 *b* alleles in sperm
0.5 *B* alleles in eggs	0.25 *BB* fertilized eggs	0.25 *Bb* fertilized eggs
0.5 *b* alleles in eggs	0.25 *Bb* fertilized eggs	0.25 *bb* fertilized eggs

Hardy-Weinberg Principle

Of course, the percentages of alleles at any chromosome locus must add up to 100% for the entire population; thus, if you know that the percentage of *B* alleles in the population gene pool is 60%, then the percentage of *b* alleles must be 40%. Putting it algebraically, $p + q = 1$, with p being the proportion of one allele and q the proportion of the other. Expanding that equation, $p^2 + 2pq + q^2 = 1$. This is the mathematical expression of the Hardy-Weinberg equilibrium principle, which shows that generation after generation, if there is random mating in an infinitely large interbreeding population, and there is no selection, migration, or mutation, the proportions of alleles and genotypes remain the same. (The principle was formulated independently in 1908 by British mathematician Godfrey Hardy [1877–1947] and German physician Wilhelm Weinberg [1862–1937].) A Punnett square shows the same result.

	p alleles in population (sperm)	*q* alleles in population (sperm)
p alleles in population (eggs)	p^2 fertilized eggs	*pq* fertilized eggs
q alleles in population (eggs)	*pq* fertilized eggs	q^2 fertilized eggs

In nature, of course, equilibrium is sometimes disturbed. Part of the population may be isolated in an area where the environment selects one or the other allele for survival; a mutation of an allele may occur; or immigration or emigration can change the balance of genotypes and alleles. Those events can change the numbers.

The Hardy-Weinberg equation can be useful for calculating the number of carriers of a trait in a given population. For instance, the trait of albinism (lack of pigmentation) is found in 0.000049 of the U.S. population, or in 1 of every 20,000 persons. The trait is known to be recessive and to be expressed (obvious) only in homozygotes. *N* is the allele for normal pigmentation, and *n* the allele for albinism. A Punnett square can be set up as follows:

	p N alleles in population (sperm)	*q n* alleles in population (sperm)
p N alleles in population (eggs)	p^2 *NN* fertilized eggs	*pq Nn* fertilized eggs
q n alleles in population (eggs)	q^2 *nn* fertilized eggs	*pq Nn* fertilized eggs

Because the proportion of albinos (*nn*, or q^2), is known to be 0.000049, q (the **frequency** of *n* alleles in the gene pool) must be q, or 0.007. $1 - 0.007 = p = 0.993$. The proportion of carriers of the trait (the heterozygotes, or 2 *pq*) is then 2(0.993)(0.007), or about 1%—a proportion that may seem surprising.[4] This is true of a great many genetic conditions that occur only rarely: Even when there is strong selection against the homozygotes, the allele continues to persist in the heterozygotes in the population.

INHERITANCE AND EVOLUTION

The Hardy-Weinberg equilibrium equation describes a theoretical condition of genetic inertia. When a population's frequencies of alleles and genotypes fail to behave according to the equation, the population is evolving for some reason.

The evolution of a group of organisms is often represented as a phylogenetic tree.

In this kind of diagram, the ancestral organism (which may be hypothetical) is shown at the base of the trunk. The various branches contain groups that have descended from the ancestor and the distance between any two groups is proportional to the theorized number of mutations (nucleotide substitutions) that have occurred in a particular gene between them over time. For example, the gene that produces the protein α-hemoglobin has been extensively studied in many animals. A phylogenetic tree based on changes in α-hemoglobin might show primitive vertebrate ancestors at the bottom. A long lower branch would end in frogs, and a shorter branch a little farther up would end in chickens. Farther up the tree, branches of varying lengths would end in mice, sheep, dogs, humans, and other mammals.

Many phylogenetic trees are constructed on the basis of anatomical similarities. These can be used as models against which hypotheses about genetic relationships are tested. Very often, genetic studies confirm the tree that was constructed from anatomical evidence, but occasionally they show surprising contradictions. A phylogenetic tree constructed with genetic data, for instance, would show birds and crocodiles closer together than either group is to lizards or snakes.

Upsetting Hardy-Weinberg Equilibrium

Several factors may lead to the evolution pictured in phylogenetic trees. They result in upsetting the Hardy-Weinberg equilibrium.[5]

Genetic Drift

The equilibrium state depends on the population being large, with mating being at random. Assume that the alleles C and c are found within a certain large population.

Under those conditions, if one parental pair produces offspring that are all CC, for example, by the laws of probability other pairs will produce some Cc and cc offspring, and the frequencies of alleles and genotypes in the gene pool will remain the same generation after generation.

Sometimes, though, a population is very small. Perhaps 10 or 20 animals of the same species become separated from others of their kind. Their initial frequencies of alleles might be in the ratio of 10 C:2 c. If selection tends to remove individuals that are Cc, the ratio might change to 10 C:1 c. (Heterozygotes would still tend to keep the c allele in the gene pool.) Chance alone can have a large effect on a small population, bringing about a change in gene frequency called genetic drift.

Mutation

If a C allele changes to become c at a certain rate per generation, and if nothing else changes, eventually a population's alleles will all be c. In practice, though, some individuals' c alleles are likely to be mutating to C at the same time, damping or canceling the $C \rightarrow c$ effect.

Nonrandom Mating

Though mating occurs at random in the Hardy-Weinberg model, it may not do so in nature. For instance, some female insects of the genotype pp may produce a certain attractive **pheromone,** whereas their PP and Pp relatives do not. Males would tend to mate with the pp females rather than with the others, which would tend to increase the p frequency in the gene pool. This type of mating, in which mates are preferentially chosen because of some genetic characteristic, is called assortative mating.

On the other hand, "opposites attract." As is often apparent in human couples, individuals may choose mates who differ greatly from them genetically. This phenomenon, which can add useful variations to their offspring's genotypes, is called disassortative mating.

Gene Flow

When individuals immigrate or emigrate at random, the equilibrium is maintained. If only organisms with a certain genotype move, however, frequencies can theoretically be disturbed. Gene flow, which refers to the movement of gene-carrying individuals from one population to another, tends to make different populations genetically similar. The more gene flow occurs, the more different alleles are introduced to the new population and the more similar the populations will become.

Natural Selection

As might be expected, natural selection is more likely than any other factor to change the frequencies of genotypes and alleles in a population. However, unlike the other forces that may lead to evolution of a population, natural selection acts on individual phenotypes; thus, it is qualitative as well as quantitative in its effects.

Individuals that are selected by the environment to leave more offspring are the fittest organisms in the population in terms of their ability to survive in that environment. Additionally, their offspring are superior to organisms with other phenotypes in their ability to survive there. Because selection acts on whole individuals, their whole genotypes are selected, not just the genes at a few loci.

Human Genes and Evolution

Luigi Cavalli-Sforza, a geneticist at Stanford University who has worked on reconstructing the history of human evolution using genetic and linguistic data from living populations, has concluded that the obvious differences among races are only superficial responses to different climates. He says that all humans are simply different populations of the same species.

Cavalli-Sforza used genetic distances between different human groups as an index of difference for classification purposes. He defines genetic distance as the difference between two populations in the frequency of a given gene. For instance, the frequency of Rh-negative individuals is 41.1% in England and 41.2% in France—a slight difference, as might be expected from the two countries' proximity. Among the Lapps, however, the frequency is only 18.7%. The genetic distance between the English and French is thus 0.1%, and that between English and Lapps 22.4%. (Other methods of calculating genetic distance are sometimes used.) He found that genetic distance did increase with geographic distance, but as a continuum; there were no sharp rises in the curve, as there would be if people belonged to separate races. Any attempt to divide humans into races on the basis of genetic distance would be arbitrary.

Using averages for different genes, Cavalli-Sforza and his colleague Anthony Edwards calculated genetic differences between populations in various pairs of continents. They found the greatest difference (24.7) was between Oceania and Africa, followed by that between Oceania and the Americas (22.6). The smallest difference, 8.9, was between Asia and the Americas, reflecting the rather recent migration from Asia to the Americas.

Starting from those data, they constructed a phylogenetic tree showing successive divisions of populations through human history. Beginning with the smallest genetic distance, between Asia and America, they began constructing the tree. Working backward to the greatest distance, between Africa and Oceania, they ended with a model of human migration beginning in Africa and moving first to Australia, then to East Asia, and finally to Europe and America. Probably the differences in genes among the various populations have arisen by both genetic drift and natural selection that occurred as people migrated.

Cavalli-Sforza feels that natural selection has acted on human culture and conscious choices as well as on human genomes. Today natural selection has probably produced a genome that will change very little, but human culture will continue to evolve. The rise of computers and expansion of communication among various groups will greatly affect human life in the future.[6]

DNA, RNA, AND PROTEIN

By the middle of the twentieth century a great deal was known about genes—what some of their chemical effects were, where they resided, how they behaved on a grand scale in populations. Yet no one had determined their chemical composition. It was not even clear whether genes were made of the DNA or the protein in chromosomes, though most biologists were converging on the idea that DNA was the genetic material. Still, the question remained of how it could be copied during cell division.

Modeling DNA

That question was answered in the early 1950s by two scientists working together at Cambridge University in England: young American geneticist James Watson (1928–) and English chemist Francis Crick (1916–). They worked out the structure of DNA by constructing a large physical model based on their knowledge of the molecule's chemistry, the method that

had been used with spectacular success by renowned American chemist Linus Pauling (1901–1994) in determining the structure of the protein α-helix.

Watson and Crick were only two of many scientists who were competing to find the structure of DNA. Pauling was another; he published a paper just before Watson and Crick, but somehow misjudged the structure, thinking it had three chains rather than two.

In Watson's entertaining account of his and Crick's work, *The Double Helix,* he described their method and his own excitement when he put together the final piece of the puzzle:

The key to [Pauling's] success was his reliance on the simple laws of structural chemistry. . . . In place of pencil and paper, the main working tools were a set of molecular models superficially resembling the toys of preschool children.

We could thus see no reason why we should not solve DNA in the same way. All we had to do was to construct a set of molecular models and begin to play—with luck, the structure would be a helix. Any other type of configuration would be much more complicated. . . .

[I] began shifting the bases in and out of various other pairing possibilities. Suddenly I became aware that an adenine-thymine pair held together by two hydrogen bonds was identical in shape to a guanine-cytosine pair held together by at least two hydrogen bonds. All the hydrogen bonds seemed to form naturally; no fudging was required to make the two types of base pairs identical in shape. . . . at lunch Francis winged into the Eagle to tell everyone within hearing distance that we had found the secret of life.[7]

Watson and Crick had found that all the previous studies of DNA could be accounted for if DNA was a double-stranded, helical molecule. The backbone of each strand was a chain of ribose (a five-carbon sugar) and phosphate groups having nitrogenous bases (A, T, C, and G)

Figure 3.8
The DNA double helix and its four bases.

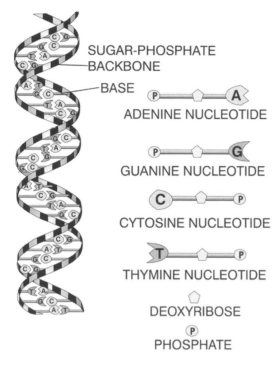

attached; the two strands were linked by hydrogen bonds that held the base pairs (A–T and C–G) together.

The **stereochemistry** of the four bases was most important for Watson and Crick's model; when they were arranged in the correct sequence, with the sugar–phosphate backbones aligned in opposite directions, the bases fit together perfectly.

Once they determined this structure, they could see that its form made possible its function of coding information and then replicating it. This was essential for explaining the transfer of hereditary data between generations and from chromosomes to cytoplasm within every cell.

Watson and Crick used an imaginative approach to determining the double-helix structure of DNA, basing the model in part on published papers, but they were aided greatly by viewing the unpublished data of physical chemist Rosalind Franklin (1920–1958), who had spent several years on producing X-ray diffraction photos of DNA. She was a methodical, brilliant worker, insisting on getting complete data before building speculative models. Franklin was working as an associate of English biologist Maurice Wilkins; the two disliked each other intensely, and Wilkins felt no special loyalty to her. He carelessly showed her photos to Watson, who immediately recognized that they confirmed his and Crick's model. That gave them the confidence to publish and take credit for solving the mystery of DNA. In 1962, Wilkins shared the Nobel Prize in Physiology and Medicine with Watson and Crick. Franklin had died of ovarian cancer four years earlier, at the age of 37.[8]

In later years, many criticized Watson and Crick for not sharing the acclaim with Franklin, and they have acknowledged that she contributed more than anyone else realized at the time. Still, not even Franklin ever claimed that the model was in part hers. Her photos showed the structure clearly, and Crick said later that Franklin had been close to the truth, but she didn't realize that the chains ran in opposite directions or that the bases were paired.[9] She probably would have come to the same conclusion in a few months. In science, however, being the first to publish accounts for nearly everything.

Like those who reached nearly the same picture of evolution that Darwin provided (see Chapter 2), Rosalind Franklin did a great deal of important work but failed to make the final intuitive leap that would have given her a major place in the history of biology. Sadly, she did not live long enough to know that later scientists gave her much of the credit for finding the structure of DNA.

After Watson and Crick published their model and it was combined with the earlier studies, the whole genetic picture that had been emerging became clear. The picture can be summarized simply as: in a one-way process, the DNA of the nucleus causes the synthesis of RNA, which causes the synthesis of protein. This statement has been called the central dogma of genetics. Though now somewhat modified, it has guided most subsequent research.

To understand the central dogma, it may be easiest to begin with RNA. Unlike DNA, RNA is usually single-stranded and nonhelical. RNA contains ribose, which is a pentose (five-carbon sugar). Four of ribose's carbon atoms, along with an atom of oxygen that looks a little like a hat in any diagram of RNA, form a ring; the fifth carbon is connected as a side branch of the ring. Hydrogen and hydroxyl (–OH) groups are bonded to the carbon atoms. (Each carbon must form four bonds, with carbons or other atoms.)

Deoxyribose is exactly like ribose, with one exception. As its name implies, it is missing one oxygen atom, because instead of a hydroxyl group on the second (2′) carbon, deoxyribose has a hydrogen atom.

Riboses can combine with phosphate (PO_4^{3-}) groups and nitrogen bases to form units called nucleotides. The nitrogen bases that are found in DNA are, as already mentioned, adenine (A), thymine (T), cytosine (C), and guanine (G). Each nucleotide unit contains just one of those bases. In RNA, there is no thymine; the base uracil (U), which has the same shape, takes its place.

The various nucleotides making up each long chain of nucleic acid are connected through their sugar and phosphate groups. A phosphate is attached to the third (3′) carbon atom of each ribose, and to the fifth (5′) carbon atom of the adjacent ribose, linking the two; a base is attached to the first carbon atom of each ribose. Chemically, the two strands of DNA run in opposite directions: A phosphate attached to a 5′ carbon is at the end of one strand, and a phosphate linked to a 3′ carbon is at the matching end of the other strand. The rest of the molecule is arranged accordingly. The two chains are held together by hydrogen bonds between the bases. The base–base links are always adenine–thymine or cytosine–guanine. The entire two-chain structure is twisted around a central axis, as a spiral staircase would be.

What happens in the reading of DNA, and the transfer of information to the cytoplasm, may be best explained with an example. Suppose part of the sequence on a DNA template is AAA TTA CTG G. In a normal person, the triplets of bases in DNA are transcribed into a corresponding sequence in messenger RNA (mRNA), the RNA that carries genetic information from the nuclear DNA to the cytoplasm. The genetic code is actually the sequence of bases in mRNA, as each base can be decoded as an amino acid; each triplet of mRNA is a codon. In our hypothetical cell, the DNA sequence is transcribed as UUU AAU GAC C in the mRNA. (Uracil takes the place of thymine in RNA.)

Before DNA can be transcribed, the molecule must be "unzipped" into the two single strands. An enzyme, DNA polymerase, moves along the DNA molecule, and the strands separate. One strand then serves as a template for copying in the 3′ to 5′ direction, and the other strand is an inactive bystander. As the enzyme moves along the DNA strand complementary nucleotides are assembled to form a single strand of RNA. The nucleotides, which are already present in the cell, are added one at a time to the 3′ end of the RNA chain. Each adenine in the DNA template is paired with a uracil in the growing RNA, each thymine with an adenine, each cytosine with a guanine, and each guanine with a cytosine. The new RNA molecule is thus complementary to the template DNA strand and identical—except for the replacement of T by U—to the inactive strand. The mRNA strand remains single.

The mRNA now becomes the source of directions for assembling proteins or polypeptides (chains of amino acids that can be combined to make proteins). It works in harmony with two other types of RNA, ribosomal (rRNA) and transfer (tRNA). The rRNA molecules make up part of the ribosomes, where translation of the code occurs.

The mRNA moves out through the nuclear membrane and travels to a ribosome, where amino acids will be assembled to make proteins as specified in the DNA code. The RNA becomes attached to the ribosome and assembles proteins, one amino acid at a time.

Ribosomes have a grooved surface into which the long mRNA molecules are fed. Each tRNA molecule has two points of attachment: One, called the anticodon, binds to a triplet on mRNA; the other is attached to a particular amino acid. In our example, the UUU AAU GAC C sequence on mRNA codes for the amino acids phenylalanine–asparagine–aspartate, because the tRNA anticodons AAA UUA CUG are linked to those amino acids.

Like transcription, translation is a three-step process—initiation, elongation, and termination—and each step depends on enzymes. Initiation begins with the attachment of rRNA to a strand of mRNA near its 5′ end. That exposes the first codon on the mRNA (known as the initiator, usually AUG). The corresponding tRNA anticodon pairs with the initiator codon, bringing its attached amino acid along. (If the initiator is AUG, the amino acid is methionine.)

During elongation, the ribosome moves along the mRNA molecule. The polypeptide grows longer, with each codon on mRNA being linked to an anticodon of tRNA that adds another amino acid to the chain. An enzyme makes it possible for the amino acids to form chemical bonds with each other, whereas anticodon–amino acid bonds in tRNA are broken, freeing the growing polypeptide chain.

Translation ends with the termination step, when a termination codon on mRNA (such as UAA) is reached. Because there are no anticodons that match these "stop" codons, no more amino acids can be added. That signals the end of the polypeptide chain, which is now free to perform its function in the cell. The unused RNA molecules can be broken down and recycled.

Sometimes a sequence is misread during transcription or translation. For instance, in our example the mRNA might miss reading the first A base, thus reading the DNA as AAT TAC TGG. The resulting mRNA would be UUA AUG ACC, coding for a completely different sequence of amino acids, leucine–methionine–threonine. This sort of mutation is called a frame shift. Sometimes a frame shift results in an abnormal protein that causes a malformation or death; or no protein at all may be formed.

The genetic codes for the 20 amino acids used in proteins and for initiation and termination of synthesis (start and stop) are shown in Table 3.2. From these amino acids, living species of organisms can form an estimated 10^{10} to 10^{12} different proteins.[10]

Though there are only 20 amino acids, 61 different triplets code for them. Because there is more than one triplet (codon) for many of the amino acids, the genetic code is called degenerate.

Besides junk DNA, there is some useful DNA within genes that may not be

Table 3.2
The Genetic Code.

Triplet of Bases	Amino Acid
UUU, UUC	phenylalanine
UUA, UUG, CUU, CUC, CUA, CUG	leucine
AUU, AUC, AUA,	isoleucine
AUG	methionine (start)
GUU, GUC, GUA, GUG	valine
UCU, UCC, UCA, UCG, AGU, AGC	serine
CCU, CCC, CCA, CCG	proline
ACU, ACC, ACA, ACG	threonine
GCU, GCC, GCA, GCG	alanine
UAU, UAC	tyrosine
CAU, CAC	histidine
CAA, CAG	glutamine
AAU, AAC	asparagine
AAA, AAG	lysine
GAU, GAC	aspartate
GAA, GAG	glutamate
UGU, UGC	cysteine
UGG	tryptophan
GGU, GGC, GGA, GGG	glycine
CGU, CGC, CGA, CGG, AGA, AGG	arginine
UAA, UAG, UGA	stop

expressed. In eukaryotic cells, most genes are split into sections. The beta-globin segment of the gene for **hemoglobin,** for instance, is split into three parts, which are separated by some noncoding DNA. The coding regions are called exons, the noncoding sections, introns. Some genes are mostly introns, and others are mostly exons.

During transcription of DNA, introns and exons are both copied into the mRNA. Before the mRNA is translated into amino acids, however, small particles of RNA and protein go to work on the mRNA. The introns are pulled out into loops, and adjacent exons are spliced together. Translation can then take place.

Geneticists think that this process is a mechanism for evolution of DNA; exons can be mixed and matched when the splicing takes place, leading to the synthesis of new proteins and new functions. Antibodies, for instance, must be made in various forms to counteract new antigens to which an organism is exposed.

Reverse Transcription

Though the central dogma was based on transcription and translation being steps in a one-way path, subsequent research showed that transcription can occur in the opposite direction also. RNA can move into the nucleus and be transcribed as DNA, which then becomes part of the cell's DNA, controlling the cell and giving it new instructions. The transcription is aided by the enzyme **reverse transcriptase.** RNA viruses that act in this way are called retroviruses.

Cancer and Genes

Some retroviruses can cause cancer when they enter nuclei of normal cells. They cause growth-promoting cellular genes that are ordinarily inactive to become active, which leads to the uncontrolled cell growth and division characteristic of cancer. Genes thus activated are called oncogenes. They produce proteins that control cell growth. Oncogenes closely resemble normal genes in the same cells.

Viruses may carry oncogenes, or they may disrupt normal functions in host cells. Viruses may also code for proteins that are needed for the viruses' own replication and that affect the host cell's regulation of its own genes. In all cases, the result is a disruption of the host cell's normal genetic activity. The disruption is copied into daughter cells as well.

At one time, the main competing theories about cancer were the mutagen theory and the viral theory. When scientists discovered that viruses contain RNA, which can cause mutations in the host cell and lead to cancer, the two theories proved to be based on different aspects of the same phenomenon.

GENETIC ENGINEERING AND BIOTECHNOLOGY

Human farmers and animal breeders have probably always chosen plants and animals for desirable characteristics and crossed them. That **artificial selection** affected the organisms' genetics but not in entirely predictable ways. (Even now, many amateur dog breeders, for example, breed animals for obvious characteristics with little idea of the underlying genetics.)

As Mendel's ideas became known, it was possible to make crosses more skillfully, with knowledge of dominant and recessive alleles. Today the chromosomal locations of many genes are known, and we have techniques for removing genes and replacing them with others. This sort of genetic micromanipulation is called genetic engineering.

For many years, the idea of genetic tinkering was restricted to science fiction, such as Aldous Huxley's *Brave New World.*[11] Today, artificial fertilization and related events are common, and we can glimpse a little of our genetic and evolutionary future. The prospect is both exhilarating and frightening, but the genie is out of the bottle. For good or ill, humans are certain to control more of our heredity in the years to come.

Genetic engineering and *biotechnology* are overlapping terms. Genetic engineering refers to making artificial changes in DNA; biotechnology is the manipulation of organisms (which may include genetic engineering) to produce useful products, such as pharmaceuticals.

Cloning

In 1996 a lamb named Dolly became the most famous sheep in the world. Ian Wilmut and other researchers in Scotland produced Dolly using DNA from an adult sheep. In their procedure, the nucleus of a cell from an adult ewe's mammary gland was implanted in another sheep's unfertilized egg, whose nucleus had been removed. The lamb was a clone (genetic duplicate) of the donor of the original mammary cell.

A clone is a population of genetically identical cells or organisms that are derived originally from a single original cell or organism by asexual methods. Or a clone can be defined as an individual that was grown from a single body cell of its parent and thus is genetically identical to it.

Plants commonly make clones, producing genetically identical plants by **asexual reproduction.** Cloning has been used in agriculture throughout history; many varieties of plants are cloned simply by obtaining cuttings of their leaves, stems, or roots and replanting them.

The body cells of adult animals and humans can be routinely cloned in the laboratory. Adult cells of various tissues, such as muscle cells, that are removed from the donor animal and maintained on a culture medium while receiving nutrients manage not only to survive but to go on dividing, producing colonies of identical descendants. By the 1950s scientists were able to clone frogs, producing identical individuals that carry the genetic characteristics of only a single parent. The technique used in the cloning of frogs consists of transplanting frog DNA, contained in the nucleus of a body cell, into an egg cell whose own genetic material has been removed. The fused cells then begin to grow and divide, just as a normal fertilized egg does, to form an embryo.

Biologists first successfully cloned mice in the 1980s, using a procedure in which the nucleus from a body cell of a mouse embryo is removed from the uterus of a pregnant mouse and transplanted into a recently fertilized egg (from another mouse) whose genetic contents have been evacuated.

Cloning a new animal from the cells of an adult (as opposed to those of an embryo) is considerably more difficult, however. Almost all of an animal's cells contain the genetic information needed to reproduce a copy of the organism. But as cells differentiate into the various tissues and organs of a developing animal, they express only that genetic information needed to reproduce their own cell type. This has tended to restrict animal cloning to the use of embryonic cells, which have not yet differentiated into blood, skin, bone, or other specialized cells, and which can more easily be induced to grow into an entire organism.

The practical applications of cloning are economically promising but philosophically unsettling. Animal breeders would welcome the chance to clone top-quality livestock. As an example of this, a black Angus bull named Bull 86 that died of old age in 1997 was naturally resistant to brucellosis, tuberculosis, and salmonellosis. Not only are those diseases dangerous for cattle, they can be passed on to humans through contamination, uncooked beef, or unpasteurized milk. A clone, Bull 86 Squared, was made from his genetic material, which had been frozen years earlier. The clone possessed all Bull 86's admirable traits, including resistance to those diseases.

Genetically engineered animals that form certain substances could be cloned in large numbers to increase the production of drugs or of human proteins that are useful in fighting disease, for example. Clones are also highly useful in biological research because of their genetic uniformity. Identical clones can be experimental and control animals.

Although the cloning of animals has many appealing aspects, it may also lead to untoward consequences. Just as **monoculture** of one kind of food plant can make a farm more vulnerable to disease or severe weather than it would be with a variety of crops, a herd of genetically identical animals may all be vulnerable to a disease or other calamity.

Many scientists object to cloning on the grounds that it is immoral or dangerous. Some of the animals already cloned have not done well; Dolly the sheep, for example, died after a few years, at about half the age at which a normal sheep would have died. A cow clone in Tennessee was found dead in a pasture for no apparent reason.

The cloning of human beings is an even more controversial subject. Already one clone has been created by a private company for the purpose of providing stem cells for research. It seems likely that other clones will follow, some intended for development into viable fetuses. If cloning can ensure the infinite replication of specific genetic traits, who would be the judge of which traits are worthy of continuation? Suppose, for example, that Adolf Hitler had had access to cloning. He would have ensured that blond, blue-eyed Aryans were cloned, and that blacks, Jews, and other groups he considered inferior were not. (A chilling movie was once based on this premise.) That may seem like an extreme example, but even benign scenarios may have hidden problems. Suppose a little boy is dying. His parents, who want to clone their dying son, may not face the fact that the child has a hereditary illness that may affect the clone as well; even if the clone survives, he may eventually pass on the faulty allele to his own children.

People like these bereaved parents may assume that the clone will have the personality of the original child, but that is unlikely. Although some personality traits are genetic, uncontrollable environmental influences—teachers, friends, foods, chemicals—also have powerful effects on a developing child. Once a unique person dies, they cannot be duplicated.

Some parents may want to use cloning to ensure that their offspring have certain traits that they think are desirable. Or they may want to be sure their children do not inherit genetic disorders that run in the family. Some of their reasons may seem excellent from any viewpoint, whereas others may be only frivolous.

Most people today are opposed to any human cloning, but occasionally someone wants to create a clone of an infertile person or of a dying spouse or child. It is probably only a matter of time until human cloning occurs successfully and results in a healthy baby. When it does, the clone will be greatly affected by its environment. Like an identical twin raised in a different household, the clone will be genetically identical to the person from whom it was cloned, but will be a unique individual.

DNA Fingerprinting

Because each person's genome is unique, the picture presented by their DNA is also unique. The use of DNA fingerprinting is now widely used. In many cases, it has been possible to prove that prisoners who had been convicted of rape or murder were actually innocent, and the prisoners have been freed.

Every human chromosome contains many repeated 15-nucleotide segments that are called **mini-satellites.** Their number of repetitions and locations can vary greatly and are unique in every person. This characteristic has proved extremely useful in forensics, making it possible to construct DNA fingerprints of suspects and compare them with DNA evidence collected at crime scenes. A rapist, for example, leaves semen in his victim. Semen, blood, and even hair contain DNA that can be analyzed.

If fragments of DNA are put in an electrical field, they will move toward the field's positive pole. Different mini-satellites will move different distances, depending on their size and other characteristics. The DNA from any individual will form a unique pattern of mini-satellites, a fingerprint.

The probability that two unrelated persons will have the same DNA fingerprint is 1 in 10 billion. People in the same family have somewhat similar patterns, but in most crimes that is not a complication.

Theoretically, a data bank of DNA fingerprints of everyone in the United States could be created for possible analysis in criminal investigations. In practice, many people object to having their DNA analyzed unless they have been accused of a crime.

Using DNA fingerprints for profiling—the construction of a person's genetic picture, complete with the potential for genetic disorders—is also possible. This is even more controversial, because it can be misused by employers, insurance companies, or others who may discriminate against someone with certain genetic characteristics.

Disease Resistance

Scientists at Texas A&M University have managed to produce a bull that is resistant to many diseases, including brucellosis, tuberculosis, and salmonellosis. The genes they used came from an animal that had natural immunity to those diseases but eventually died of old age. Its cells were used for producing a clone that had the same characteristics, including immunity.

Salmonellosis can cause diarrhea, fever, and cramps and is sometimes fatal. Brucellosis may lead to brain infections and heart dysfunction, and tuberculosis is a deadly respiratory ailment. An animal that is naturally resistant to the diseases would be safer for humans to eat, yet would not have to be given antibiotics as preventive medicine. If it were also vaccinated, it would then be made as much as 100 to 1,000 times as resistant to the diseases as other cattle are.

Gene Therapy

Genetic engineering of human embryos can potentially make it possible to insert "good" alleles into nuclei and to remove "bad" alleles, providing natural resistance to diseases.

One promising study was carried out in dogs, which sometimes have a disease called Sly syndrome. It is caused by a deficiency in an enzyme that breaks down certain sugar molecules and leads to an inability to walk, cardiovascular disease, and corneal clouding. Newborn dogs affected with the syndrome were injected with a retrovirus encoding the enzyme; they began producing the enzyme and remained healthy and mobile for over a year. Control dogs not given the retrovirus developed the syndrome within six months. The same method that protected the puppies could be used to prevent diseases of the blood and liver.[12]

Crohn's disease, which causes painful inflammation and ulcers, may be caused by a defective gene. In this disease, the immune system attacks the normal bacterial flora in the gut; apparently a mutant form of the gene *Nod-2* causes the body's normal controls to go awry. The normal gene encodes a protein that recognizes the bacteria belonging in the gut and prevents the immune system from attacking them. Crohn's patients produce a modified protein that cannot perform that function. Now that the defective gene has been identified, it may be possible to substitute the normal form and protect people from the disease.[13]

Another disease that may be amenable to genetic engineering is Parkinson's, which causes neural degeneration leading to tremors, very slow movements, and other symptoms. The genetic aspect has not been established, but a few families do show clear inheritance of Parkinson's. They have a mutated form of one of three genes that produce proteins that normally work together and break down a substance called isoform. If one gene is a mutant, the isoform builds up in the cell and leads to degeneration. Substituting the normal form of the affected gene could protect those families.[14]

Drugs and Vaccines

The smallpox virus once was as much of a threat as HIV is today, causing a disease that disfigured or even killed its victims. Because a vaccine was developed in the eighteenth century and was gradually made available to people all over the world, the disease was virtually eliminated. (There have been a few recent reports of cases in India.) The last confirmed outbreak was in 1977 in Somalia. In that year, the World Health Organization declared the disease eradicated. Since then the vaccine has not been given, and few people today are immune to smallpox.

However, some scientists recently created a deadly version of mouse smallpox in the laboratory by accident. Lethal human viruses could be created in a similar way, either accidentally by scientists or deliberately by terrorists.

The researchers, who were trying to develop an effective contraceptive vaccine, made a simple change in a mousepox virus, inserting a gene that produces a body chemical called interleukin-4 (IL-4). They hoped to stimulate an immune reaction against mouse eggs, which would produce a contraceptive effect. Instead, the gene acted to suppress the part of the immune system that normally fights off viral infection. All the animals given the altered virus died within days, though ordinarily mousepox causes only mild illness in mice.

Some smallpox viruses have been saved in the United States and Russia for research purposes, and it is still possible to make vaccines from them. If someone inserted the gene for human IL-4 in the human smallpox virus, they could produce a far more virulent strain that could even be resistant to vaccines. Because the majority of people on Earth today have not been vaccinated against smallpox, even the ordinary strain could pose great danger if it fell into the hands of terrorists.

Antibiotics, which are produced by molds, can be altered by genetic engineering of the molds' DNA. New types of antibiotics to replace older ones (to which microbes have become resistant) can be produced in large quantities.

Monoclonal Antibodies

In the 1980s a Chicago-area man was diagnosed with cancer that appeared to be incurable. Chemotherapy put the cancer into remission but made the patient sick and weak. When the cancer returned a few years ago, instead of chemotherapy he chose to try an experimental drug. The drug is made of specially designed proteins, known as monoclonal antibodies, that are attached to radioactive isotopes. The antibodies delivered their radioactive freight to the tumors and killed the cells. His cancer disappeared after only one treatment and has stayed away since then. Unfortunately, not all results with these drugs have been so good, but monoclonal antibodies remain promising.

Antibodies are the body's first line of defense against bacteria or viruses, recognizing and attacking foreign proteins (antigens) on their surfaces. In 1975, scientists developed a way to make highly uniform antibodies against a specific antigen. They had great potential as a way of destroying diseased cells while leaving healthy ones undamaged and minimizing side effects.

To prepare the antibodies, drug companies use a technique developed in 1975 by immunologists César Milstein (1927–2002) and Georges Köhler (1946–1995) at the Laboratory of Molecular Biology in Cambridge, England. The antigen is injected into a mouse, which generates thousands of antibodies to it. The blood cells that make the antibodies are removed and combined with tumor cells that live indefinitely (in contrast to the blood cells, which would soon die outside the body). The combined cells are called hybridomas. Because the antibody-producing hybridomas all are from the clone of one blood cell, they are called monoclonal.

The first monoclonal antibodies did not perform well as drugs. Their main drawback was being made of mouse proteins, which were rejected by the human immune system.

Later, however, genetic engineering techniques made it possible to graft the mouse antibody onto a human antibody and lower the immune response. Still later, human genes were substituted for the mouse antibody-producing genes, producing an entirely human antibody.

There still are many drawbacks to the use of monoclonal antibodies as drugs; they are too large to enter cells and can't be given orally because (like all proteins) they are destroyed by stomach acids. Also, they appear no more effective than chemotherapy against cancer. Because they have fewer side effects, they may be most useful when used in combination with chemotherapy.[15]

Enzymes

Because enzymes are proteins, which must be encoded by DNA, enzymes are an obvious target for genetic engineering. Both the enzymes that the cells themselves use and those that may be useful for human purposes can be controlled and even mass-produced. Producing beer, baked goods, and many other foods involves enzymes that can be engineered for flavor and texture.

One instance of genetic engineering for human uses is the manufacture of vegetarian cheese. Making cheese from milk depends on curdling the milk with the enzyme chymosin, which comes from calves' stomachs.

To produce a vegetarian cheese that tastes much like the real thing, manufacturers insert the genes for calf chymosin into a yeast. The chymosin produced is thus from a nonanimal source and can be used with a soy substitute for milk.

Many enzymes are now sold for use in washing clothes in cold water or with very small amounts of detergent. The enzymes break down the proteins in many stains. The bacteria *Bacillus licheniformis* and other bacteria have been engineered to produce effective enzymes for laundry.

Engineering Plants

Since about 1980, many agricultural products of genetic engineering have appeared. They have been developed to resist frost, to be less susceptible to disease, to produce their own natural insecticides. Rice, corn, strawberries, and many other crops have become genetically engineered foods.

Although genetically modified plants have been a boon to agriculture, they have become very controversial (see Chapter 6).

Mapping the Human Genome

The idea of mapping the entire human genome was first seriously discussed by scientists in 1984. Over the next five years, much opposition emerged from both scientists and the public, based mainly on arguments that the work would require too much research time and money and would have the potential for dire ethical, legal, and social consequences. Eventually, however, the project began. In sharp contrast to Mendel working alone in his garden and to Watson and Crick building a model with toylike components, the project required contributions from hundreds of scientists in many countries working with computers.

Sequencing is the determination of the number and order of components in DNA or proteins. For DNA, the components are

TOOLS AND TECHNIQUES: GENETIC ENGINEERING

Even in an undisturbed cell nucleus, a certain amount of snipping and splicing of DNA segments take place. With the growth in knowledge about the makeup and locations of various genes, scientists have learned how to do the snipping and splicing deliberately to modify an organism's genome.

The critical steps in genetic engineering are similar to those in a word-processing program—cut, copy, and paste. As you might expect, the ubiquitous compounds enzymes carry out these operations. Restriction enzymes cut; polymerases copy; and ligases paste.

Restriction enzymes, which are found only in bacteria, cut a DNA molecule at a certain point. The enzyme EcoRI, for instance (made by *Escherichia coli* bacteria), searches for a GA-containing sequence and cuts the DNA between G and A. It leaves the rest of the DNA molecule unscathed. Some restriction enzymes cut DNA unevenly, leaving "sticky ends," or short sequences of single-stranded DNA, that are useful for later pasting.

A geneticist can take the cut portion of DNA and insert it in a **vector** for transfer to another cell. The vector is likely to be a plasmid from a microbe. (A plasmid is a DNA fragment that is not part of the chromosomes. Plasmids are more common in bacteria than in eukaryotes.) The DNA, with its sticky ends, is added to the plasmid, along with a restriction enzyme that attacks the plasmid's DNA. DNA ligase is added to weld the sticky ends of the cut gene fragment and of the plasmid together, which creates a recombinant plasmid.

The recombinant plasmid next carries the gene fragment to the host microbe. Not all cells will take up the plasmid, but some will (and these can be found with a chemical **marker**). Because the host microbe now thinks the recombinant DNA is part of its genome, it replicates that DNA along with its own, with the aid of the DNA polymerase enzyme.

Different individuals, even those closely related, have different DNA sequences. So, one individual might have a GATTCGA sequence, and another might have GATTGACGA. A restriction enzyme that snipped DNA between G and A would cut the first sequence in two places, the second in three places, yielding DNA fragments of different lengths. These are known as restriction fragment length polymorphisms, or RFLPs (pronounced **rif**-lips). Polymorphism differences between human populations have been useful in studies of human evolution and migration (Cavalli-Sforza 2000).

Polymerases can be harnessed for expanding the amount of DNA available for study, for use as a medicine or commercial product, or for DNA fingerprinting. In the DNA polymerase chain reaction (PCR), the polymerase first makes a copy from a DNA template, then copies the two strands, and so on. At each round of copying, all the existing strands are copied, and so there are soon millions of strands of identical DNA.

---

nucleotides; for proteins, they are amino acids. Two sequencing projects—the International Human Genome Sequencing Consortium and the private-sector group at Celera Genomics—raced to be first to publish the entire sequence, and the race ended in a tie. The two groups published their results at the same time, in February 2001. The publication of the draft human genome sequence generated tremendous excitement, with the directors of both groups overjoyed at the accomplishment and what it would mean for biology. Dr. Francis S. Collins of the consortium said "One doesn't want to get carried away, but I have to say I'm pretty carried away. . . . I'd put this on the short list of big moments in biology." Dr. J. Craig Venter of Celera stated that "it

Figure 3.9
Human chromosomes 4 and 7, showing the locations of some genes for diseases.

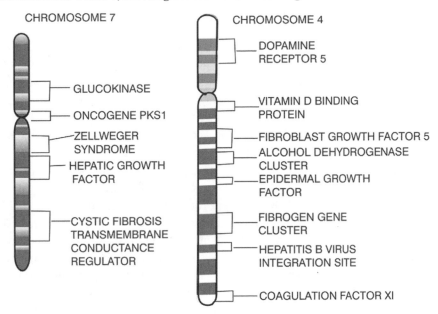

will change science and our view of ourselves as we go forward."[16]

David Baltimore (the American virologist who shared a Nobel Prize for Physiology or Medicine for codiscovering reverse transcriptase) added, "I've seen a lot of exciting biology emerge over the past 40 years. But chills still ran down my spine when I first read the paper that describes the outline of our genome."[17]

Before completion of the entire genome sequence, biologists had thought there might be hundreds of thousands of human genes. However, the total number of protein-coding genes appears to be only about 30,000—only one-third more than in a nematode worm. This surprising result may have enormous effects on scientists' understanding of human evolution as well as genetics.

Medicine in the Postgenomic Era

Humans have a gene that codes for a certain protein enzyme, known as phenylalanine hydroxylase (PAH). (Recently it has been renamed phenylalanine 4-monooxygenase, but the PAH name is still commonly used.) PAH converts the amino acid phenylalanine to tyrosine. Because phenylalanine and some of its products are toxic to developing nervous systems, PAH is important for preventing mental retardation in newborns and children.

The disorder called phenylketonuria (PKU), inherited as an autosomal recessive, results when a newborn baby's cells cannot make the PAH enzyme that breaks down the amino acid phenylalanine. Phenylketone and other **toxins** build up in the urine and blood, and the nervous system may be irreparably harmed. At this time treatment consists of a diet low in phenylalanine. Eventually, however, it may be possible to use genetic engineering to replace the defective gene with a normal one. It is just one promise of new genetic knowledge and techniques.

In the excitement over determination of the human genome, some physicians and geneticists may be overoptimistic about the ramifications for medicine. Shortly after the announcement that the human genome had been mapped, the *British Medical Journal* published interviews with three prominent specialists about their predictions for the future. Their reactions were mixed; one was cautiously enthusiastic, one visionary, one skeptical.

John Burn, director of the Northern Genetics Service at Newcastle University, is positive. He feels that the National Health Service offers unique opportunities to conduct clinical research and develop comprehensive services. He also welcomes what he describes as "the endorsement of clinical geneticists as the leaders in the move to bring genetics out of specialized centers into everyday practice." Burn has created a multidisciplinary genetics unit that has brought together many different specialists for working with patients.

Also the director of the Imperial Cancer Research Fund's clinical cancer genetics network, Burn hopes to prevent some breast, ovarian, and colon cancer. He commented,

About 5% (1 in 20) of patients have a single gene defect underlying their cancer but we can only find them after there have been multiple cancers in the family. As mutation scanning becomes faster and cheaper we will be able to offer this testing to all 60,000 patients each year. By focusing on finding the 1 in 20 whose families are at greatest risk and where prevention is possible for example by regular removal of bowel polyps in those who are predisposed to bowel cancer we will demonstrate the benefits of genetic testing and reassure the skeptical.

Another positive view was offered by Gordon Duff, professor of molecular medicine at the University of Sheffield. Duff sees the new knowledge of human genetics as an invaluable tool for providing preventive medicine and controlling the spiraling cost of medical care. This would be an advantage not only for patients in wealthier countries but also for those in poorer countries with their huge burdens of infectious disease.

Duff sees **genomics** taking more than 25 years to "change the way we do medicine," but he predicts that pharmacogenomics will become a standard part of some fields within 3 to 5 years. (Pharmacogenomics is the use of an individual's genetic information to determine how they will respond to drugs.) In addition, he expects the food industry to invest in **nutritional genomics.** It is important, for instance, to determine the optimum daily dose of vitamin A for pregnant patients, and the daily individual requirement could be determined genetically. Some of Duff's colleagues are working on that connection.

Duff also is excited about using screening tests for genetic susceptibility to diseases. Screening would make it possible to modify risks by prescribing preventive medicines, adjusting the diet, choosing the best work environment, and making changes in lifestyle. In addition, people would be better equipped to manage their own health.

He even forecasts tomorrow's consumers dropping into their local retail pharmacy for genome profiling. They can obtain advice and then download more information from the Internet about how to modify the risk of developing a disease to which they are especially susceptible.

The rosy picture painted by Burns and Duff was greeted with skepticism by the third interview subject, Neil Holtzman, director of genetics and public policy at Johns Hopkins University. Holtzman thinks many people are being persuaded by "genohype."

It is not that he plays down the advances in understanding the genetic basis of disease, for he acknowledges that the identification of more than 1,000 single gene disorders that affect 1–4% of the population may well lead to reliable predictive tests and improved therapy. What Holtzman opposes is exaggerating the importance of genetic factors, which may cause people to forget the need to clean up the environment and tackle socioeconomic contributions to illness.

In many common diseases, the outcome is determined by complex genetic, environmental, and behavioral interactions. In Holtzman's view, "it will be difficult, if not impossible, to find the genes involved or develop useful and reliable predictive tests for them."[18]

Chapter 4 explains how genetic traits are expressed, directing embryonic development and adult functions.

NOTES

1. N. Angier, "Researchers Locate Gene that Triggers Huntington's Illness," in *The Science Times Book of Genetics,* ed. N. Wade (New York: Lyons Press, 1998).

2. Gregor Mendel, "Experiments in Plant Hybridization," read at the meetings of February 8 and March 8, 1865 (published in 1866 in *Proceedings of the Natural History Society* Brünn). Available online at www.netspace.org. mendelweb.

3. N. Wade, "Y Chromosome Bears Witness to Story of the Jewish Diaspora," *New York Times,* May 9, 2000.

4. D. L. Murray and J. A. Bond, *An Experience with Populations* (Reading, Mass.: Addison-Wesley, 1971).

5. W. A. Jensen et al., *Biology* (Belmont, CA: Wadsworth, 1979).

6. L. L. Cavalli-Sforza, *Genes, People, and Language* (New York: North Point Press, 2000).

7. J. D. Watson, *The Double Helix* (New York: Atheneum, 1968).

8. B. Maddox, *Rosalind Franklin: The Dark Lady of DNA* (New York: HarperCollins, 2002).

9. F. Crick, "The Double Helix: A Personal View," *Nature* 248 (1974): 766–69.

10. A. L. Lehninger, *Biochemistry* (New York: Worth, 1972).

11. Aldous Huxley, *Brave New World* (New York: Harper & Row, 1946).

12. K. P. Ponder, et al., "Therapeutic Neonatal Hepatic Gene Therapy in Mucopolysaccharidosis VII Dogs," *Proceedings of the National Academy of Sciences USA* 99 (2003): 13102.

13. J.-P. Hugot et al., "Association of Nod2 Leucine-Rich Repeat Variant with Susceptibility to Crohn's Disease," *Nature* 411 (2001).

14. H. Shimura et al., "Ubiquitination of a New Form of Alpha-Synuclein by Parkin from Human Brain," *Science* 293 (2001): 263.

15. A. Pollack, "The Birth, Death and Rebirth of a Novel Disease-Fighting Tool," *Washington Post,* October 3, 2000.

16. N. Angier, "Genome Shows Evolution Has an Eye for Hyperbole," *New York Times,* February 13, 2001, p. D1.

17. *Nature* 409 (2001): 814–5.

18. "Bringing Ordinary Doctors into the Genetics Party," *British Medical Journal,* 322 (April 28, 2001): 1016.

Organs and Systems

This chapter summarizes information about animal organs and systems. All multicellular organisms have specialized structures for functions, such as getting and using food and oxygen, reproducing, ridding themselves of wastes, and so on. Because of the extensiveness of the topic, the focus here is on human anatomy.

Until the sixteenth century, knowledge of human anatomy was based largely on studies of other animals conducted by Galen (129–c. 216). Galen was a Greek physician and teacher whose writings dictated the theory and practice of medicine for about 1,500 years after his death. Born in Greece but doing most of his work in Rome—where the Roman religion prohibited dissecting human corpses—Galen had to base his studies on monkeys, dogs, goats, and other animals. Those observations were fruitful: Galen described seven pairs of cranial nerves, the valves of the heart, and the differences in structure between arteries and veins. In physiology, he was the first to show that the arteries carry blood, not air. Galen's descriptions of some human organs were erroneous, however; even his description of the human uterus was based on his dissection of a dog. Unlike the human uterus, a dog's is Y-shaped.

For several hundred years, medical students were taught from Galen's books, not from the human body. That all changed when young Flemish physician Andreas Vesalius (1514–1564) was appointed a lecturer in surgery at the University of Paris, with the responsibility of giving anatomical demonstrations. Vesalius became suspicious of the textbooks written by Galen and began using his own approach to learning and teaching anatomy—doing dissections himself rather than having his teaching assistants do them, learning body structure from cadavers, and reevaluating ancient teachings. Vesalius wrote his own complete textbook of human anatomy (*De humani corporis fabrica libri septem,* commonly known as the *Fabrica*), which was printed in 1543, and went to Venice to supervise the artists who illustrated it. The elaborate drawings were probably made in the studio of the great Renaissance artist Titian.

The classic book is still remarkable today for its detailed illustrations of muscles and other parts of the body, many

APPLICATION TO EVERYDAY LIFE FEATURE: THE STEM CELL DEBATE

Throughout the history of Western medicine, most medical treatment has been limited to using drugs or performing surgery. Today a new era is beginning in which it may be possible to grow new, healthy cells and replace damaged cells. Whether this revolution in medicine will occur depends partly on politics and religion.

Whether or not to carry out research on stem cells derived from human embryos has become a divisive issue for Americans, and has created some unexpected alliances. Opposition comes from antiabortion groups who think it wrong to use embryonic tissue for this purpose, because even the smallest embryo has the potential to develop into a human being. Some ethicists are opposed, too, because they think viewing an embryo that is fated to die as a source of cells might even pave the way toward viewing fatally ill patients or prisoners on death row as sources of cells. Support comes from scientists and politicians who say the embryonic stem cells will be invaluable for research and possible therapeutic use in patients with Parkinson's disease, Alzheimer's disease, paralysis from spinal cord injuries, diabetes, cancer, and other disorders.

The embryos used for stem cell research would never have become babies in any case: During the in vitro fertilization procedure (in which eggs are fertilized in the laboratory before being implanted in a woman's uterus), many embryos are created in excess of those wanted, and the extra ones must be discarded. Scientists hope to use these for research. In addition to the embryonic cells formed as a by-product of in vitro fertilization, some embryos have also been created for the specific purpose of providing stem cells.

Antiabortionists prefer confining research to the stem cells that are derived from adult cells, such as those in the placenta and bone marrow. On the other hand, biologists such as Nobel laureate Paul Berg—one of the pioneers in the field of recombinant DNA technology—say that adult stem cells are rare and hard to recover, in contrast to the easily available embryonic cells. The blood in umbilical cords (discarded following births) is another possible source of stem cells and does not pose the same ethical dilemmas that embryonic cells do.

Why do stem cells have so much value? It is because they are undifferentiated cells that have the potential to differentiate into a variety of cell types, such as brain or endocrine gland cells. Cell differentiation in complex multicellular organisms occurs during embryonic development and is regulated through the expression of various genes. The first cells formed from division of a fertilized egg are the undifferentiated stem cells that have caused so much controversy. Because all those cells have the diploid number of chromosomes, they all have the potential to develop into any kind of cell. As the cells multiply, they also change, becoming more specialized, but in the early stages of development, they retain some ability to be transformed into more than one type of cell. The last cells to be formed in the embryo are the most specialized and the most resistant to change. The cells are gradually transformed by chemicals from their surroundings during embryonic development. Theoretically, at least, stem cells can be transplanted to adult tissues, where they will become like the surrounding adult cells. This has led to exciting possibilities for treating patients with damaged tissues.

Stem cells have proved therapeutic in some studies of rodents. In one study, injured adult rats were assigned to two groups. Group A received transplants of stem cells from embryos; group B did not. The animals in group B could not support their body weight or move their back legs in a useful way. Those in group A regained some useful movement in their legs and could partially support their body weight. Other studies of rodents showed that transplants aided animals having disorders similar to Parkinson's disease, multiple

CONTINUED

sclerosis, and Lou Gehrig's disease (amyotrophic lateral sclerosis) in humans ("Stem Cell Transplants" 2001).

In a study of mouse embryonic stem cells, the cells were cultured in a special process that caused them to resemble the insulin-producing cells of the pancreas. The cells even produced small amounts of insulin. This research is far from solving the problem of diabetes (in which mammals produce too little insulin in response to sugar), but it is an important contribution to the research on that condition (Vogel 2001).

In 2001, President George W. Bush announced that he would allow federal taxpayer money to be used for research into stem cells from human embryos. The research would be limited, however, to stem cell colonies, or lines, that had already been established. The government would not support the destruction of new embryos.

Though some scientists doubted that many such lines were available, the National Institutes of Health (NIH) found there were about 60. Some were in Australia, Israel, and other countries outside the United States.

Neither those who supported stem cell research nor those opposed to it were completely satisfied with the president's announcement; on the other hand, each group felt he had somewhat supported their view (Seelye 2001).

Sources: Seelye, K. Q. 2001. "Bush Backs Federal Funding for Some Stem Cell Research." *New York Times,* 10 August.
"Stem Cell Transplants." June 2001. *Brain Briefings* (Society for Neuroscience).
Vogel, Gretchen. 2001. "Stem Cells Are Coaxed to Produce Insulin." *Science* 292: p. 615.

having background landscapes with details such as the gallows that furnished many of Vesalius's study cadavers. It was the most accurate and thorough picture of human anatomy that had ever been presented.

Because of Vesalius, anatomy became a scientific discipline, affecting medicine, physiology, and other fields of biology. Even today, his pragmatic attitude toward studying dead bodies to benefit living patients may help inform our own deliberations about appropriate subjects for medical research and practice.

FROM CELLS TO ORGAN SYSTEMS

Anatomy can be studied on many levels, from cells to systems of organs. Specialized cells make up tissues—for example, neurons make up neural tissue, and muscle cells make up muscle tissue.

Tissues of different types make up organs with certain functions—circulatory, muscle, and neural tissues make up the heart. Finally, organs are the components of systems—the stomach, pancreas, and other organs make up the human digestive system. The human body is an exceedingly complex combination of systems, organs, tissues, and cells specialized for their different functions.

Multicellularity

A single-celled organism can handle all its needs for obtaining, transforming, transporting, releasing, and eliminating matter and energy; but the cells of multicellular organisms are specialized in different ways. More organized than one-celled organisms, multicellular living things have a great variety of cells with specific functions. These cells must be produced from the simple cells resulting from division of the fertilized egg.

The change from amorphous blobs to functioning cells—that is, the differentiation of cells—is important at many times in the lives of living things. It occurs not only during embryonic development of all multicellular organisms but also during metamorphosis in insects and in the development of young plants that have asexually split off from parents. In adult organisms, growth of new tissues and regeneration of lost parts are made possible by cell differentiation. Differentiation is a one-way street; once specialized, a cell does not become unspecialized again.

Differentiation

As an undifferentiated group of cells becomes specialized and immutable, it goes through a series of stages. In the beginning, a fertilized egg has the capability of becoming any cell in the organism. It divides several times without growing larger, producing cells called blastomeres. Each blastomere contains genetic instructions about its "address"; that is, whether it will be located anteriorly or posteriorly in the developing organism. Those blastomeres are thus slightly differentiated—their options have been limited somewhat.

The Hox Genes

A group of genes called the Hox genes direct the process of assigning blastomeres to their addresses in the developing organism. In all mammals, clusters of Hox genes are arranged in sequence on four different chromosomes; there are 38 Hox genes in all. (They are arranged sequentially in invertebrates also, but there are fewer genes.) As the cluster toward one end of a nucleus becomes active, it is translated and transcribed to produce proteins that direct the cell to become part of the anterior part of the organism. Other cells are activated to become, in turn, parts of the trunk and tail

region. The brain is differentiated earlier than other organs, the genitals later. So the pattern of Hox genes on the chromosomes becomes a rough guide for the final form of the body itself. In a human embryo, the entire design of the body is laid out within two weeks of conception.

It appears that the body is built in segments from anterior to posterior, with each segment modified from the one anterior to it. An analogy in word processing might be: create segment 1, copy and paste to create segment 2, modify segment 2, copy and paste to create segment 3, and so on. As in word processing, this procedure would be more efficient than creating each segment independently.

When mutations in Hox genes are brought about in the laboratory (by X-ray irradiation of the embryo's parent), the genes are reprogrammed, causing body parts to grow in the wrong areas: in *Drosophila*, long legs rather than antennae may appear on top of the head, for instance. Geneticists named these mutations homeotic (meaning similar).

When researchers were studying this sequential development of similar segments and the genes responsible for regulating them, they found each of the Hox genes contained an identical sequence of 183 base pairs, which they called the homeobox. The developmental genes enveloping the homeobox were named Hox.

The proteins made by Hox genes clamp onto the chromosomes, activating a series of subordinate genes that bring about a cascade of effects on development.

All animals, including humans, have Hox genes, but no fungi or plants have them. That makes the genes very helpful for classifying any organism.[1] In modern classification methods, phylogenetic trees are based more on genetic similarities than on superficial resemblances. If an organism has no Hox genes, its position on the family tree that represents all life is far removed from the branches that represent animals.

As the final result of differentiation in humans, several systems of organs have been created. These cooperate to carry out important functions in the body: using and distributing materials, shielding and supporting the body, communicating throughout the body, and creating a new individual.

USING AND DISTRIBUTING MATERIALS IN THE BODY

Living organisms are continually undergoing building-up (anabolic) and breaking-down (catabolic) processes. In sum, these processes are known as metabolism. Although metabolism is similar in general ways, the details differ in different organisms.

When food enters an animal body, it is broken down in a series of steps by the digestive system, then absorbed into the circulatory system, where it is carried by blood and lymph. The nutrients in food are carried to cells in all parts of the body, where they can be used for tissue growth and repair and for energy.

Air enters the body through the respiratory system. The oxygen moves across membranes to the circulatory system and is distributed to cells. Within cells, nutrients and oxygen enter the TCA cycle (see Chapter 1), in which the nutrients are used for materials and energy, and CO_2 and water are formed. Wastes of metabolic processes, dissolved in water, are removed from the body by the excretory system.

Moving Materials across Membranes

Many molecules move through air or water by the process of diffusion, the natural tendency of molecules to move from an area where they are more concentrated to an area where they are less concentrated. When diffusion occurs across a membrane, it is called osmosis. Osmosis is a more complex process than diffusion, because the membrane can be permeable to some molecules but not to others. Whenever a molecule passes from one cell to another— to say nothing of moving from one tissue or organ to another—it must contend with the problem of transportation across semipermeable membranes. For instance, cell membranes have special proteins that transport certain protein molecules from one side of the membrane to the other.

Respiratory System

Many athletes prepare for competition by training in the mountains of Colorado. Because the atmospheric pressure is lower there, the athletes get less of the oxygen their bodies need. Their bodies adapt to the conditions in Colorado by increasing the number of red blood cells (which carry oxygen), and the adaptation continues for a while afterward, even when the athletes are at lower altitudes. This presumably increases their endurance.

All animals need oxygen for metabolizing food, as discussed in Chapter 1, and they get it from the air. Air brought into an organism by the **respiratory system** has a higher percentage of O_2 (21%) than of CO_2 (0.03%). The oxygen passes into the blood, where it is carried by **hemoglobin** to all parts of the body. As it is used by cells, oxygen is combined with carbon atoms to become CO_2. The CO_2 then leaves the body, through either the respiratory system or the **excretory system.**

Fish and other aquatic animals absorb O_2 from their environment when the water passes over their gills, respiratory structures that are rich in blood vessels. The O_2 diffuses into the gills' blood and is dispersed through the animal's body. Because the oxygen must be dissolved in water, gills are useful only for aquatic animals.

When animals evolved and moved onto land, one of the problems to be solved was that of obtaining O_2 and losing CO_2. Amphibians developed simple lungs where some gas exchange could take place. Their moist skin is the major site for gas exchange, however. Reptiles also have simple lungs, but depend on them more than amphibians do.

Evolution and Respiration

Later evolution led to more elaborate lungs. In humans, the lungs have several lobes, composed of hundreds of small, thin pouches called alveoli. Each alveolus is encircled with tiny blood vessels, capillaries, that join to form venules which lead to the pulmonary veins.

Air enters the respiratory system through the external nares (nostrils) and passes through the pharynx, larynx, trachea, and bronchi. The bronchi branch into bronchioles, which branch further and further until the air passages end in the lungs' alveoli. All the organs together make up the respiratory system.

Oxygen in the Blood

The respiratory system provides the points of entry and exit for O_2 and CO_2. After O_2 leaves the lungs by diffusing into the alveolar capillaries, it enters the pulmonary veins that lead to the heart. The heart pumps the freshly oxygenated blood through arteries to all parts of the body. Cells absorb the O_2 for use in cellular respiration, during which CO_2 is produced. The CO_2 diffuses into the blood.

In the blood, O_2 is carried inside red blood cells (RBCs) by the red pigment hemoglobin (which becomes oxyhemoglobin). While O_2 is transported within red blood cells, CO_2 is carried in the plasma (the fluid portion of blood), mainly in the form of HCO_3^- (bicarbonate) ions. The reversible chemical reactions

$$H_2O + CO_2 \rightleftharpoons H_2CO_3 \text{ (carbonic acid)} \rightleftharpoons H^+ \text{ (hydrogen ion)} + HCO_3^-$$

occur continually, shifting in one direction or another to keep the pH and the ions within normal ranges.

The blood returning to the heart through the veins has an ever increasing amount of CO_2 in proportion to O_2. When this high-CO_2 blood reaches the heart, it is pumped through the pulmonary arteries to the lungs. There it can pick up a fresh supply of O_2, lose its load of CO_2, and begin a new cycle.

Because the brain can go only a few minutes without a supply of oxygen, healthy lungs are of vital importance. Lung tissue is delicate and can be injured by disease, smoking, pollution, and other causes.

There are several syndromes involving the lungs of infants. One is respiratory distress syndrome (hyaline membrane disease), which almost always occurs in prematurely born infants. The lungs are stiff and the alveoli close off, diminishing absorption of oxygen into the blood. Though the condition is often fatal, it can be prevented or treated.

Adults may develop such conditions as lung cancer or emphysema. In lung cancer, there is a **malignant** tumor in the lungs; it can spread, or metastasize, to other organs. Emphysema is characterized by abnormal, permanent enlargement of some alveoli, accompanied by the destruction of their walls, and without obvious fibrosis (formation of scar tissue). As the alveoli lose their elasticity, they cannot function normally. Smoking is often the cause of lung cancer and emphysema. In addition, the bronchi and lungs can be infected by bacteria or

Figure 4.1
Human Respiratory System.

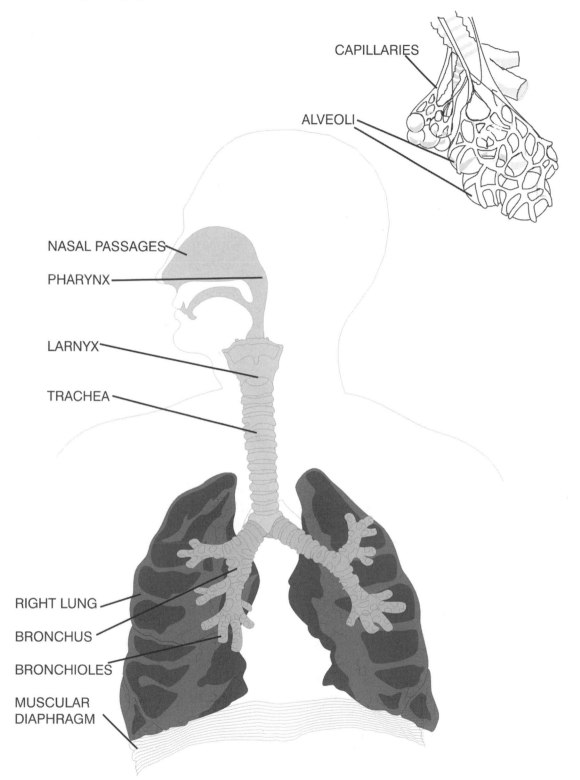

CAPILLARIES

ALVEOLI

NASAL PASSAGES

PHARYNX

LARNYX

TRACHEA

RIGHT LUNG

BRONCHUS

BRONCHIOLES

MUSCULAR
DIAPHRAGM

viruses, with bronchitis or pneumonia being the result.

Recently researchers at universities in Michigan and Texas developed a device that might aid or even replace failing lungs in humans; they have tested it successfully in animals. Because lungs can now be transplanted, the device will most likely be used in patients waiting for transplants.

Circulatory System

For about 40 years, medical researchers have linked high levels of **cholesterol** in the blood to increased risk of cardiovascular disorders, such as heart attack and stroke. Though cholesterol is found normally in the body and plays an important role in some structures, a high level of it is a danger signal.

Now a new suspect is emerging: A high level of homocysteine, a substance that builds up when the amino acid methionine is not sufficiently **metabolized,** is also linked to cardiovascular disorders (such as heart attack, stroke, peripheral vascular disease, and **venous** blood clots). The jury is still out on how dangerous homocysteine is, but there is an easy way to reduce it—eating less meat (which contains methionine) and taking supplements of three of the B vitamins (folic acid, B_6, and B_{12}). Because this therapy is harmless and inexpensive, people do not have to wait years for research results before trying to lower their own homocysteine levels.[2]

Why is it important to determine whether cholesterol or homocysteine impedes circulation? Heart disease and other disorders of the circulatory system are the major cause of death in North America, causing more than two of every five deaths in the United States. The circulatory system pumps gases, food, and water—along with important enzymes and hormones needed by various organs—to all parts of the body. It also carries waste materials, such as carbon dioxide and urea, away from cells and takes them to the liver for **detoxification** and to the kidneys for excretion.

In small, simple animals, there are no heart, veins, or arteries. Blood sloshes around their bodies through large **lacunae** and blood vessels.

Fish have a two-chambered heart containing one auricle and one ventricle. Blood enters the auricle and is pumped out by the ventricle. With the movement onto land and the development of lungs, more elaborate hearts were needed to collect and dispense oxygenated and unoxygenated blood. Amphibians have a three-chambered heart; reptiles, birds, and mammals, four chambers.

The Human Circulatory System

In the human heart, blood from the lungs enters the left atrium through the pulmonary veins, bearing fresh oxygen. The blood passes through valves (which prevent the blood from flowing back) into the left ventricle. This is the most muscular part of the heart. It pumps blood out into the aorta, the massive artery that courses down through the body and gives off branches to the head and limbs. The aorta and other large arteries branch repeatedly, ending in arterioles that branch again into microscopic capillaries. Capillaries are barely large enough for blood cells to move through them. Other capillaries also act to collect CO_2 and waste materials from cells, and send them toward the heart through a series of venules and veins. One large vein, the hepatic portal vein, enters the liver, where much waste is removed from it—the breakdown products of alcohol, for example. The hepatic vein then collects blood from the liver and carries it toward the heart. Two large veins, the anterior and posterior venae cavae, enter the right atrium of the heart, bringing unoxygenated blood from the entire body. This blood passes through a valve into the right ventricle,

Figure 4.2
Human Heart.

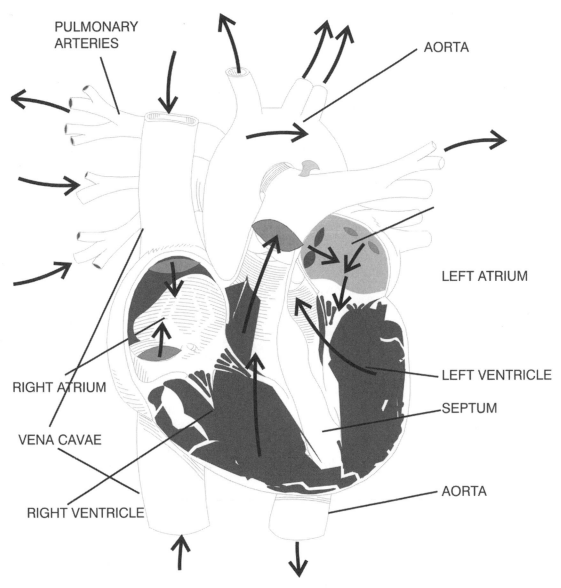

PULMONARY
ARTERIES

AORTA

LEFT ATRIUM

RIGHT ATRIUM

LEFT VENTRICLE

SEPTUM

VENA CAVAE

AORTA

RIGHT VENTRICLE

which pumps it out through the pulmonary arteries to the lungs. There it can absorb oxygen and again return to the left atrium.

Blood Pressure

The arteries that carry blood from the heart to other organs have strong, thick walls compared with those of veins and capillaries. Arterial walls have large amounts of elastic tissue and muscle, enabling arteries to withstand the pressure placed on them as the heart pumps blood through them.

The pressure exerted against arterial walls by the blood is blood pressure. Blood pressure readings consist of two numbers, such as 120/75. The first number—the systolic pressure—is higher because the heart muscle is contracting and forcing the blood out. The second number—the diastolic pressure—is lower because the heart is relaxing and the ventricle is refilling with blood. When it contracts, an artery narrows its own diameter and raises the blood pressure. When it relaxes, the diameter becomes greater and the blood pressure drops. A typical normal blood pressure reading is 110/70.

High blood pressure, or hypertension, is the condition in which the blood's pressure against arterial walls is much higher than normal. Hypertension is associated with several disorders, including heart disease. If a person with hypertension also has **arteriosclerosis,** the blood pressure may force a clot or piece of artery wall to enter the bloodstream, where it may block circulation to vital organs.

The capillaries extend near every cell in the body. Because their cell membranes are only one cell layer thick, oxygen and other materials can easily move from the blood into the surrounding **tissue fluid;** from there it passes through cell membranes into various cells.

Carbon dioxide and other wastes pass in the opposite direction, from cells to tissue fluid to capillaries. These capillaries join to form larger vessels that eventually form venules and veins. Veins are relatively larger and thinner-walled than arteries, and the blood pressure is lower in veins.

Lymph

The lymph system is made up of a network of vessels that move lymph—tissue fluid that is within the lymphatic system—around the body and empty it into some veins in the chest. Once inside the veins, the lymph is the plasma portion of blood. Lymph carries the monoglycerides and fatty acids produced by digestion of fats.

Lymph nodes scattered around the lymph system are clumps of tissue that filter dead cells, bacteria, and other foreign materials from the lymph. White blood cells (leukocytes), which can scavenge bacteria, are produced in lymph nodes. Because the nodes are involved in fighting infection, they may become enlarged in anyone who is diseased.

One of the important functions of the circulatory system is maintaining various kinds of homeostasis (equilibrium). By bringing food and oxygen to cells, it makes obtaining energy and materials possible; it regulates body temperature; it maintains a constant pH of the tissue fluid (enzymes can only act within certain pH ranges); and by removing wastes, it protects the cells from being poisoned.

Digestive System

In recent years, obesity has become a major U.S. public health concern as well as a matter of personal distress because it is associated with several chronic diseases. Cardiovascular illnesses and diabetes, especially, are brought about or worsened by being overweight. The problem is rapidly getting worse, as the American way of life becomes more sedentary.

Based on phone surveys of adults conducted by the Centers for Disease Control and Prevention and state health departments from 1991 to 1998, the prevalence of obesity—defined as a body mass index (BMI) of at least 30—increased from 12.0% in 1991 to 17.9% in 1998. (BMI equals a person's weight in kilograms divided by the square of height in meters. A person who is 5 feet, 7 inches high and weighs 140 pounds would have a BMI of 22.) The increase occurred in all states; in both sexes; in all age groups, races, and educational levels; and among both smokers and nonsmokers. The percentage of obesity increased most in 18- to 29-year-olds (7.1% to 12.1%), those with some college education (10.6% to 17.8%), and Hispanics (11.6% to 20.8%).

Because the participants probably tended to underestimate their weight and overestimate their height, the prevalence of obesity shown in this study is likely to be a conservative estimate. Since then, the percentage has risen even more, to 22.3%. A variety of factors lead to obesity, but in most cases it results from taking in more calories than are used up in daily activity.[3]

Obesity is especially common among the urban poor. For a family struggling to pay the rent and utilities, it may not be possible to buy the fresh produce and lean meat that help in weight control. Also, youngsters who live in neighborhoods plagued with drug dealers and other crime may stay inside to be safe, rather than getting outdoor exercise. A recent study at Stanford University showed that at all economic levels, children who watch television for long periods tend to weigh more than those who watch less television. Not only are they exercising less, but they also are continually exposed to commercials for junk food that is fattening.[4]

According to an Agriculture Department report, about 28% of blacks have poor diets (which are especially lacking in fruits and vegetables), whereas 16% of whites have poor diets. Obesity is the most obvious result of too much food, but being overweight is associated with other ills also. Type 2 diabetes and sleep apnea, and perhaps asthma, are all linked to obesity.

People who must buy cheap food are likely to buy things like macaroni and cheese, corn muffin mix, and white bread—all containing some nutrients but high in carbohydrates. They may fry foods, which makes them more fattening, rather than steaming or broiling them. Blacks and Hispanics also tend to buy meats that are high in fat, which are less expensive and may be ingredients in recipes traditional in their cultures.[5]

Whatever we eat, and however much of it, enters the digestive system. In this system food is broken down physically into very small pieces, then altered chemically. Very large molecules—proteins, fats, and starches—in hamburgers and milkshakes first become **proteoses, disaccharides,** and other large molecules. These are further broken down into small units, such as amino acids, simple sugars, and fatty acids. Most of these are absorbed through the intestinal wall into the blood vessels that surround the intestine. Unused materials that stay in the digestive system instead of being absorbed continue moving through the system (which is one long, curved tube) and are lost in the feces (ejected).

The human digestive system has evolved from much simpler systems and is well adapted to our **omnivorous** way of life. Humans have what is called a "tube within a tube" body plan, based on what began in invertebrates as a straight digestive tube that stretches from one end to the other of a cylindrical organism.

Preparing Food for Digestion

Food taken into the mouth is first chewed (masticated) by the teeth. Humans have four kinds of teeth: On each side of the upper and lower jaws are two incisors, one canine, two premolars, and three molars. (The back molars, the "wisdom teeth," are last to erupt and may never appear.) These give us a useful built-in tool kit. Incisors allow us to slice meat, canines to spear it. Premolars and molars are large grinding teeth that help us eat tough stalks of plants.

Salivary **glands** in the mouth secrete enzymes. These break down starchy foods by **hydrolysis.**

Having been reduced in size by the teeth and chemically changed by enzymes, food passes back over the tongue into the pharynx; then it moves down a tubular structure, the esophagus; and enters the stomach. The stomach has a very low pH because the **gastric juices** have a high content of hydrochloric acid, which further breaks down the **bolus** of food. (The stomach itself is protected from the acid by a layer of mucus.) Large protein molecules are split by the acid into proteoses and peptones and those into polypeptides and eventually amino acids. Bacteria in food are also destroyed in the stomach.

Digestion

After leaving the stomach, the food enters the duodenum, where most **digestion** occurs. This is the first part of the small intestine. Ducts from the gallbladder and the pancreas enter here, draining their secretions into the digestive tract. Bile (produced in the liver and stored in the gallbladder) emulsifies fats. The pancreatic enzymes break down starch and other compounds. Because pancreatic juice contains sodium bicarbonate, it is basic; it neutralizes the acidic material arriving from the stomach.

The food—now unrecognizable—enters the next portion of the small intestine. Starch and sugars are broken down into simpler sugars.

Absorption

Amino acids and simple sugars are absorbed through the **villi** of the intestinal wall into the many capillaries in the wall and go on to the rest of the circulatory system. Fatty acids, the breakdown products, have a different fate: They are absorbed by **lacteals** leading to the **lymphatic system.** They are repackaged into chylomicrons (protein-coated droplets) that travel in the lymph to the point in the chest where it empties into the circulatory system.

Most food has now been broken down and absorbed, but some will be lost from the body. Water has also been absorbed into the blood. **Fiber,** such as that in plant stems, is the main component of the food that enters the colon, or large intestine. A large part of the colon's contents are the intestinal bacteria, such as *Escherichia coli.* These bacteria are usually harmless—some even synthesize the vitamins biotin and K—but their presence in water or other material is a signal that fecal contamination has taken place, a useful indicator for public health studies. The last part of the colon, the rectum, ends at the anus.

Nutrition

In passing through the digestive system, the foods that enter the body as mixtures of substances are broken down to nutrients that can be absorbed and used by cells for energy and materials. The three main classes of nutrients are proteins, fats, and carbohydrates. From these the body can rearrange atoms and construct the proteins, fats, carbohydrates, and nucleic acids it needs for growth and reproduction. Other substances in food are **vitamins** and **minerals.** Though the body uses only small amounts of vitamins and minerals, they are important participants in many chemical reactions.

PROTEINS

Proteins are large compounds made of amino acid units. Every amino acid has the basic structure

$$H_2N-CH-COOH$$
$$|$$
$$R$$

in which R is a side chain that differs and distinguishes one amino acid from others.

Two amino acids can become linked together as a peptide:

$$H_2N-CH-CO-HN-CH-COOH$$
$$\qquad | \qquad\qquad\quad |$$
$$\qquad R \qquad\qquad\quad R$$

The bond between carbon and nitrogen in a peptide linkage (CONH, or CO–HN) is called a peptide bond. Long chains of 50 or more amino acids, linked by peptide bonds, make up proteins.

The shape of a protein determines what it can do, and each protein twists and folds into a definite three-dimensional shape. Because there are more than 10,000 kinds of proteins in the human body—each with a specific sequence of amino acids and three-dimensional shape—they can perform an enormous number of different jobs in the body. Proteins catalyze chemical reactions, transport compounds, regulate processes, and help form structural materials in the body. Enzymes and hormones are proteins. In addition, muscle tissue and cell parts are made of protein.

Proteins broken down during digestion to amino acids move into the bloodstream by absorption and circulate throughout the body. The amino acids may be remade into proteins as needed or used for energy. If they are present in excess of the body's needs, their amino (NH_2) groups are removed and excreted in **urea.**

Proteins come from both animal and plant sources. Digestion of proteins yields amino acids. Twenty common amino acids are found in natural proteins, and nine of them—"essential" amino acids—must be supplied by foods. The others, called nonessential amino acids, are synthesized by body cells. Many animal proteins contain all nine essential amino acids, but plant proteins usually lack two of them. In spite of that, a diet consisting only of plant foods

Table 4.1
The Amino Acids in Natural Proteins.

Essential	Nonessential
Lysine	Alanine
Phenylalanine	Arginine
Leucine	Asparagine
Threonine	Aspartic acid
Methionine	Cysteine
Isoleucine	Glutamine
Tryptophan	Glutamic acid
Histidine	Glycine
Valine	Proline
	Serine
	Tyrosine

can be healthful, because some plant proteins lack certain amino acids and others lack different ones. Vegetarian diets can provide all the essential amino acids if foods are chosen carefully to complement each other. Rice and beans, for instance, are incomplete in themselves but can provide all the essential amino acids if they are eaten at the same meal.

CARBOHYDRATES

Carbohydrates include starches and sugars. Most are made entirely of C, H, and O atoms and stored energy. The simple sugar glucose (the product of photosynthesis) and its relatives, such as **fructose** and **lactose,** are an important part of our diet. The table sugar we use is sucrose, a disaccharide made up of fructose and glucose units. Sucrose is found in honey and sugarcane; most of the sucrose we use is refined from sugarcane and sugar beets. Among the starches we eat are those in potatoes and rice. In the digestive system, starches are broken down to the sugar units of which they are composed. In cellular respiration, each turn of the TCA cycle stores energy from glucose in the phosphate bonds of ATP, which can later be used for releasing energy as needed for muscular contraction and chemical reactions.

We all know carbohydrates as energy foods. Students cramming for exams or putting in late-night hours on their computers have long known the craving for candy, chips, and cookies that accompanies those sessions, but they may have assumed the food served only to keep them awake and upright. Now there is evidence that thinking actually drains glucose from a key part of the brain.

A team of scientists at the University of Virginia carried out research on rats and found that increasing the animals' memory load lowered the glucose level as much as 30% in the brain area concerned with memory for location. (The rats were making their way through a maze.) The effect was more pronounced in older rats, and their brains also took longer to recover normal glucose levels.

In a follow-up study of older rats, glucose levels in the active brain areas dropped by 48% during the maze task and did not return to normal until 30 minutes later. In contrast, recovery occurred immediately in younger rats. The rats' performance could be boosted by glucose injections. The researchers pointed out that their study may have important implications for human learning and nutrition and for age-related problems with learning and memory.

Within limits, carbohydrates are obviously important to the body. Fiber, for example, is an indigestible substance that helps in the formation of soft stools that can be eliminated easily. Fiber also slows the intestine's absorption of glucose, lessening the surge in the blood sugar level that follows eating. It may play a role in lowering blood cholesterol and lowering the risk of colon cancer as well.

The problem with carbohydrates is that most of us are eating too many of them while exercising too little to use the energy and materials they provide. Excess glucose is converted to fat, which is stored in deposits under the skin and around internal organs, such as the heart.

The fat deposits can impede blood circulation while increasing the demand on the heart and lungs. Cholesterol and other substances associated with fat can form plaques that narrow the arteries. If the affected arteries are the **coronary arteries,** the heart muscle itself may be deprived of what it needs. A heart attack can result. If they are arteries to the brain, a stroke can occur. Either a heart attack or a stroke can be fatal or cause enormous permanent damage to the body.

FATS AND OILS

Like carbohydrates, fats are needed by our body, but only in small amounts. Although some of the fat stored in our bodies began as the sugar and starches in food, some actually began as **fat.** The "well-marbled" beef that has been prized traditionally as being the highest quality is riddled with fat; unlike cheaper cuts of beef having large amounts of visible fat, marbled fat cannot easily be separated into lean and fat sections. All fried foods contain some fat or oil. Fats and oils are the solid and liquid forms of the class of chemicals called lipids; lipids include **triglycerides, fatty acids, phospholipids,** and **sterols.**

Nearly all the lipids we eat are triglycerides. Fat in our bodies also consists of triglycerides. Because triglycerides are too large to pass through cell membranes, they must be broken down to **monoglycerides** and fatty acids during digestion.

Some foods of animal origin contain sterols, such as cholesterol. Plant foods contain other sterols but do not supply any cholesterol. Though cholesterol has attained a bad reputation because of its association with heart disease, it is a necessary compound in the body; it is used for making **bile acids,** a vitamin, and several hormones.

The main wrongdoer in heart disease is actually low-density lipoprotein (LDL), a compound that promotes the formation of plaques in blood vessels. Plaques may block the coronary blood vessels, causing a heart attack. On the other hand, high-density lipoprotein (HDL) clears cholesterol from the blood. For that reason, laboratory reports of blood lipids include the HDL-to-LDL ratio as well as the total cholesterol level. A high proportion of HDL decreases the risk of heart attack. A desirable level of total blood cholesterol is less than 200 mg/dl; a level of 240 or higher is considered high. A level between 200 and 240 may require attention, especially if the HDL level is low (less than 35) and the LDL level is high (160 or more).

For many people, eating less **saturated fat** and cholesterol raises the level of HDL and lowers LDL. Dietary fiber also lowers LDL. Other factors interact with these to affect a person's risk of heart disease. For some people at risk, the cholesterol-lowering drugs called statins are helpful. They inhibit an enzyme that is needed for the body's control of cholesterol levels.

The complex paths in which fats, proteins, and carbohydrates are metabolized have many intersections. For that reason, any of the nutrients may be converted to other types during metabolism. For example, protein is sometimes considered a non-fattening food, but when eaten in excess of the body's needs, some of the atoms in protein can be rearranged to form fat, with the nitrogen becoming part of urea and excreted.

VITAMINS AND MINERALS

Metabolic reactions are assisted by enzymes, some of which need the presence of **coenzymes.** Some vitamins—notably the B vitamins—act as coenzymes, affecting the metabolism of glucose, triglycerides, and amino acids.

Pellagra, a disease once common in rural areas of the United States, is seldom seen in this country today. It brought on lesions of the skin and mucous membranes, severe gastrointestinal symptoms, and mental aberrations. Pellagra is caused by a deficiency of niacin (one of the B vitamins), which can be obtained from fish, meats, tomato juice, peanuts, and mushrooms. People with high protein intakes never have a deficiency of niacin, because the amino acid tryptophan can be converted to niacin in the body. In contrast, people whose diets contain a great deal of corn are in danger of developing pellagra, because corn is low in both niacin and tryptophan.

Niacin, thiamin, riboflavin, biotin, pantothenic acid, vitamin B_6, folic acid (folate), and vitamin B_{12} are the B vitamins, all of which act as coenzymes. They make it possible for the body to carry out vital reactions that produce energy, synthesize and break down amino acids and fats, and synthesize glucose from noncarbohydrate compounds.

All of the B vitamins are water-soluble, meaning that cooking foods by boiling them can dissolve the vitamins in the cooking water. They also can be destroyed by light, leaching, alkalinity, and chemicals. Microwaving or steaming foods will protect the active vitamins from destruction.

Another water-soluble vitamin is vitamin C, which helps the immune system and has other functions in the body. It is also known by the name of ascorbic acid.

The B and C vitamins make up one of the two major groups of vitamins, the

Table 4.2
Vitamins in Human Nutrition.

Vitamin	Food Sources	Function in Human Body
A	Yellow and green vegetables, dairy foods	Supports reproduction, needed for vision, promotes cell differentiation, bone growth, and health of mucous membranes and skin
B_1 (thiamin)	Liver, whole grains, legumes, pork	Regulates reactions that remove CO_2
B_2 (riboflavin)	Meat, fish, dairy foods	Needed for extracting energy from food
B_3 (niacin)	Liver, whole grains, legumes, lean meat	Needed for extracting energy from food
B_6 (pyridoxine)	Vegetables, whole grains, meat	Needed for using amino acids
Folic acid (folate)	Green vegetables, whole wheat, legumes	Needed for using amino acids and nucleic acids
B_{12}	Eggs, dairy products, muscle meats	Needed for using amino acids and nucleic acids
C (ascorbic acid)	Citrus fruits, tomatoes, green peppers	Needed for building bone, teeth, and cartilage
D	Eggs, dairy foods, fortified milk, cod liver oil (also synthesized by the body after exposure to sunlight)	Needed for growth and for depositing calcium and other minerals in bones
E	Leafy green vegetables	Prevents damage to cell membranes
K	Leafy green vegetables	Needed for clotting blood

water-soluble group. The other major group is the fat-soluble vitamins, including vitamins A, D, E, and K. The two groups differ in how the body absorbs, transports, and stores them: Water-soluble vitamins are absorbed directly into the blood from the intestine; fat-soluble vitamins enter the lymphatic system first. Water-soluble vitamins float in the bloodstream; fat-soluble vitamins are carried in **chylomicrons,** like triglycerides. Water-soluble vitamins are not stored in the body but are excreted; fat-soluble vitamins are stored in the liver and in fat and are not excreted. As a result, water-soluble vitamins must be taken in

often but are unlikely to cause toxicity. Fat-soluble vitamins, in contrast, can build up to toxic levels if taken in excess.

Because fat-soluble vitamins are toxic in either too small or too large amounts, it is important to get enough of the vitamins—especially from a balanced diet—but to avoid overdosage. Vitamin A, for instance, may be deficient in people who eat few vegetables, are alcoholics, have liver disease, or cannot absorb fats normally. They may develop night blindness, **xerophthalmia,** or even total blindness.

If a person stores too much vitamin A, it may lead to birth defects or damage to the

Table 4.3
Minerals in Human Nutrition.

Mineral	Food Sources	Function in Human Body
Calcium (Ca)	Dairy foods, dark green vegetables	Helps build or maintain bones, teeth, blood, nerves
Phosphorus (P)	Dairy foods, meat, poultry, grains	Helps build or maintain bones and teeth; helps regulate acid-base balance
Sulfur (S)	Proteinaceous foods	Makes up part of tendons, cartilage, and muscle
Potassium (K)	Milk, meat, fruits	Helps regulate acid-base balance, body balance, and nerve function
Chlorine (Cl)	Table salt, dairy foods, fish	Helps form digestive enzymes; helps regulate acid-base balance
Sodium (Na)	Table salt, dairy foods, fish	Helps regulate acid-base balance, body water balance, blood pressure, and nerve function
Magnesium (Mg)	Leafy, green vegetables, whole grains	Activates enzymes for protein synthesis
Iron (Fe)	Eggs, legumes, lean meats, whole grains, leafy green vegetables	Needed for synthesizing hemoglobin in red blood cells; used in extracting energy from food
Iodine (I)	Saltwater fish, dairy foods, vegetables, iodized salt	Makes up part of thyroid gland products
Water (H_2O)	Drinking water, liquids	Transports materials, takes part in synthesis and hydrolysis reactions, regulates body temperature

bones or liver. Vitamin A itself is in animal foods, such as egg yolks and liver, or it can be made from certain precursors in plants, such as the β-carotene in sweet potatoes and carrots. The current recommended **daily value (DV)** for vitamin A is 5,000 international units (equivalent to 3,000 μg [micrograms] of β-carotene). Toxicity is unlikely unless 6 to 10 times the DV is taken as supplements or the person uses certain acne drugs that are made from vitamin A derivatives.

Minerals are inorganic solid materials that are needed for formation and function-ing of various body parts and compounds. Though it is not a mineral, water is listed in Table 4.3 because it is an inorganic substance that is needed by the body.

Although U.S. diets tend to be high in junk foods, they are often low in milk and other sources of calcium. One of the most important minerals—and the one most likely to be neglected—is calcium. In healthy adults, new bone formation takes up about 400 mg calcium per day, about the amount in the circulating blood. That means we must take in more calcium every day to replace what is used. Adults need at least 800 mg calcium per day, and adolescents need 1,000–1,400 mg. A cup of milk or yogurt contains 350 mg calcium, and fortified products have even more. However, women and girls who drink soda pop in place of milk may not get as much calcium as they need. Some dieters think milk has too many calories, but many low-calorie dairy products are available.

A widespread genetic irregularity makes it hard for some people to digest the **lactose** in dairy foods because they lack the proper enzyme. (This problem is common in African Americans, Native Americans, and some other populations. It is rare in northern Europeans.) They can meet their calcium needs with calcium-fortified foods, such as orange juice, or with calcium supplements. Some good sources of calcium are broccoli, some dark green leafy vegetables, tofu (if made with calcium), canned fish (eaten with bones), and fortified bread and cereal products. Although nutritionists recommend getting as much calcium as possible from foods, calcium supplements in the form of calcium citrate or calcium carbonate can also help provide the recommended daily allowances.

With all the sources of calcium available, some people might wonder if they are getting too much calcium. As much as 2,000 mg per day seems to be safe for most, but anyone at risk for kidney stones should discuss calcium with their doctor and perhaps ingest less calcium, because most kidney stones are made of calcium compounds.

Excretory System

As more and more Americans' kidneys fail because of damage from high blood pressure, diabetes, or other conditions, dialysis—the use of a mechanical kidneylike device—becomes more necessary. Without this procedure, which mimics the normal excretory process, many dialysis patients would die within a few days.

Either dialysis or normal excretion helps keep the ranges of salts and water in the tissue fluid normal. The excretory system also eliminates cellular waste products from the body. These products include salts, carbon dioxide, nitrogenous wastes, and sometimes water. (Intestinal wastes consist mostly of fiber and bacteria. They are eliminated by the digestive system and do not enter the tissue fluid.) Homeostasis between an organism and its environment is largely maintained by the excretory system.

Nitrogenous Wastes

Nitrogenous wastes may be in one of three forms—ammonia, urea, or uric acid—depending on the organism and the environment in which it finds itself. Ammonia is **soluble** in water, as is urea. Ammonia is extremely poisonous. When it is converted to urea, however, it can travel in the bloodstream. The least soluble nitrogenous waste is uric acid, the pastelike substance excreted by birds, reptiles, and insects. Because little water is needed for it, many of the organisms in dry environments excrete wastes as uric acid.

In single-celled organisms and coelenterates, excretion of wastes consists of simple diffusion. This is sufficient because most of these organisms are surrounded with salt water and have an interior environment that is much like the exterior one. More complex animals have ducts that collect wastes in one organ, where salts and water can be treated selectively to maintain the proper balances. That organ, the kidney, may be simple or complex, again depending on the animal itself and its surroundings. In humans, the lungs, skin, and liver aid in excretion.

Some organisms that live in or near salt water have special adaptations to rid themselves of salt, which they need to do to avoid becoming **dehydrated.** Large sea turtles excrete salty tears; some seabirds have salt glands in their heads that drip solutions; even some marsh plants can excrete salt from glands in their leaves. Other organisms have adaptations that allow them to live in fresh water without taking in too much water by diffusion.

Let's follow the progress of some compounds through the human excretory system. These compounds are toxic in excess amounts and must be eliminated. They pass from the cell into a capillary and move in the bloodstream.

The Human Kidney

When the compounds reach one of the two kidneys in a **renal artery,** the artery branches repeatedly until it forms many knotlike structures surrounded by capsules. A capillary–capsule combination is called a nephron; this is the kidney's functional unit. Blood pressure **filters** plasma into the capsule, along with suspended wastes, salts, and glucose. Large protein molecules and cells in the blood stay in the capillary. The filtrate in the capsule is similar to plasma at this point.

Some large molecules that cannot move into the filtrate by diffusion are moved by **active transport,** in a **secretion** process. H^+ ions also are moved into the filtrate, which helps maintain the normal pH range of the blood. The filtrate continues moving from the capsule into a tubule; as it does, glucose, water, and some salts are **reabsorbed** into the surrounding capillaries. By the time the filtrate passes to the end of the tubule, it consists of water and waste products alone and is called urine.

Millions of collecting tubules merge to form a ureter from each kidney to the single bladder. One duct, the urethra, passes out of the body from the bladder.

Other Methods of Excretion

The liver also participates in excretion. Any amino acids used for energy may be broken down to pyruvic acid (which enters the TCA cycle) and to ammonia (NH_3) in the liver. The liver converts the ammonia to urea, which enters the bloodstream and is carried to the kidneys. In addition, the liver breaks down red blood cells, toxins, drugs, and excess enzymes and hormones.

The excretion process is vital because of the toxicity of many waste products and other substances. If disease or injury damage the kidneys and decrease the amount of excretion, an artificial kidney can be used for dialysis of a patient. During dialysis, the blood (or **peritoneal fluid**) is routed through a machine that filters it, much as a real kidney does. Vital compounds and water are returned to the fluid before it reenters the patient. The patient may temporarily lose several pounds of water during each dialysis, which must be carried out for several hours every week.

Though the products of immune reactions against bacteria may be removed by excretion, bacteria themselves are too large to be eliminated that way. They must first be attacked by mechanisms of the immune system.

SHIELDING AND SUPPORTING THE BODY

Vertebrates, including humans, have no external skeleton for support and protection. Instead, our internal skeleton supports us and provides attachment points for muscles. Muscles contract and pull on the bones, making it possible for us to move around.

We have a strong, flexible skin that contains the body and provides a defense against microbes and injury. Additional protection against microbes, **allergens,** and so on is furnished by the immune system.

Integumentary System

Many of us try to achieve a "healthy" tan by staying out in the sun too long, not using sunscreen, and otherwise making ourselves vulnerable to skin cancer. Skin cancer is the most common form of carcinoma (cancer) in the United States. The three major types are basal cell carcinoma, squamous cell carcinoma, and malignant melanoma. Although the first two are easily curable, the third is extremely dangerous and can be fatal. Even basal cell and squamous cell carcinomas can cause great damage and disfigurement if left untreated. Malignant melanoma diagnosed early can usually be cured but if diagnosed later may spread and cause death.

Obviously, then, avoiding skin cancer is important. Aside from the danger of affecting other vital organs, the skin itself is important to health.

The integumentary system includes the skin and the various kinds of tissues embedded in it, such as sense receptors of the nervous system, small muscles for erecting hair, and exocrine glands. All of those tissues make the skin much more than just a bag holding the body together. It is our interface with the environment and our first line of defense against infection.

In humans, skin is even more important for warmth and protection than in other vertebrates, because humans have comparatively little hair on their bodies.

A variety of sense receptors reside in the skin, especially in areas that are important for receiving stimuli from the environment, such as the lips and fingertips.

Skin Bacteria

The skin is a habitat for large numbers of bacteria, fungi, and yeasts, the normal microbiota (sometimes called the normal flora). Skin is moist and warm, providing a hospitable environment for bacteria and other microorganisms. Indeed, many bacteria have become so adapted to conditions on human skin that they can grow only under those conditions.

Though we need to keep the natural flora within bounds by bathing, the antibacterial campaign may backfire if it becomes too aggressive. Most natural flora are harmless and may help repel **pathogens** by competing with them for resources. In fact, many bacteria found in humans are beneficial and are required for a healthy life. The common skin bacteria *Staphylococcus alba,* for example, is usually harmless and may protect us from overly large numbers of *Staphylococcus aureus,* which is responsible for many infections.

Because humans must spend energy to maintain our body temperature near 37°C (98.6°F), the skin performs an important function by helping conserve heat. Skin also protects our aqueous internal environment from drying by evaporation.

Small glands in the skin secrete sweat or oil. Sweating (perspiring) cools the body by using body heat to evaporate the sweat. Oil keeps the skin pliable and less likely to get cracks through which bacteria may enter.

Epithelial Tissue

Epithelial tissue makes up skin, nails, and hair. It also lines the digestive tract, urinary tract, and other tracts that have external openings.

Skin color depends on pigments in the epithelial cells, such as melanin. Melanin is produced when the skin receives the ultraviolet (UV) rays of sunlight, providing a tanned or freckled barrier against damage from UV light.

The health of the skin reflects the conditions of other body systems. Although melanoma is the most threatening skin condition, many others may also affect health or cause disfigurement. Poor blood circulation can result in pale or bluish skin that does not heal easily if cut. The skin of diabetics often suffers from the lack of insulin that all cells need to metabolize sugar for carrying out cell functions.

Most teenagers have a mild form of acne; this can become serious and may also occur in the adult years. Rosacea somewhat resembles acne but is not related to it. Usually found in middle-aged adults, rosacea causes flushing, acnelike bumps on the skin and discomfort. Bacterial infections can cause boils or other symptoms. Dermatitis, or skin irritation, can result from exposure to certain chemical compounds or even from psychological stress. Lice, mites, and other parasites may find homes on the skin. Red patches on the skin, covered with silvery gray scaly areas, are symptoms of psoriasis.

Skin was an innovation in the vertebrates, whose skeleton is buried within the body. Invertebrates have no need of skin for protection, being surrounded by exoskeletons, **cuticles,** or shells. Vertebrates have internal endoskeletons that protect internal organs only.

Immune System

Acquired Immune Deficiency Syndrome (AIDS) attacks the body's natural defenses against other diseases, leaving it vulnerable to damage or destruction by them. AIDS is caused by the human immunodeficiency virus (HIV), a retrovirus that enters lymphocytes and merges with their DNA. The DNA is then transformed by the HIV so that it makes copies of the HIV rather than of itself; the HIV copies burst out of the cell and attack new host cells.

Some success in controlling AIDS has been achieved with a class of anti-HIV drugs called protease inhibitors, which block a protease part of HIV. When protease is blocked, HIV makes copies of itself, but the copies can't infect new cells. At first these drugs seemed to be a miracle cure for AIDS, but their effects can fade over time. Each new HIV virus produced may be slightly mutated, making it resistant to a drug that was effective against the older type of protease.

Until protease inhibitors and other drugs that somewhat controlled the virus were developed, the disease was usually fatal. Even now AIDS is a grave illness requiring constant medication and monitoring. A vaccine containing DNA has been used with some success to protect macaque monkeys against the simian version of HIV. It may be possible to use a similar technique to develop vaccines for human HIV. For now, prevention—avoidance of risky sexual practices, of contaminated hypodermic needles, and of infected blood and other body fluids—is the most effective strategy against AIDS.

The fight against AIDS has focused attention on the **immune system** as never before. Although AIDS specifically attacks lymphocytes, immunity involves other parts of the body as well.

Our bodies fight disease-causing microbes on three levels. First, we are adapted to keep them from entering our tissues at all. Second, certain cells act as scavengers that can remove bacteria from tissue fluid. Third, on a molecular level we have **antibodies** that can tie up bacterial **antigens** chemically, making it harder for them to do any damage. If the body is successfully protected against disease, we have achieved immunity.

Levels of Protection

On the first level, the epithelial tissue keeps most microbes on its surface. Unless skin is broken, bacteria cannot penetrate it. Inside the digestive tract (which is actually a tube through the body and outside the tissues) more epithelial tissue secretes hydrochloric acid and other substances in which bacteria cannot grow. Mucus in the digestive and respiratory tracts, as well as earwax, carry bacteria toward the body's surface.

On the second level, we have white blood cells, or leukocytes. Several types of white blood cells are in our bloodstream, each with a distinct shape. During an infection, the number of white blood cells and proportion of one type may rise greatly; different types respond to different bacteria. (This characteristic is often useful for physicians trying to diagnose an illness by examining blood with a microscope.) The white blood cells travel in the blood and lymph—even pass through cell membranes into tissues—where they can kill bacteria by surrounding and swallowing them.

Finally, microbes that get through those two barriers may be vulnerable to attack by antibodies produced by certain white blood cells called lymphocytes. Bacteria and viruses have protein molecules on their surfaces that act as antigens (foreign substances in the body). Antibodies are Y-shaped molecules produced by lymphocytes; the ends of the antibody "arms" connect with the antigens, taking them out of circulation or making it impossible for the microbes to make toxins. When an antigen enters the tissues, a lymphocyte having specific antibodies for that antigen is stimulated to produce more lymphocytes having those antibodies. Many antigen–antibody reactions ensue, making an immune reaction; if they are completely successful, the infection is cured. The effects of an immune reaction may be very obvious, such as swelling, pus formation, or redness.

Antigens and Antibodies

How do the lymphocytes know what antibodies to make against the millions of different possible antigens?

During embryonic development, some cells that are differentiated in the liver, spleen, and thymus become lymphocytes. Some of them move to lymph nodes and the bone, where they continue to form throughout life. The antigens formed by the developing body are seen as "self" proteins by the developing lymphocytes, and normally no antibodies are formed against them. Other antigens, however, stimulate the production of specific antibodies against them.

The production of antibodies is the basis of vaccination for diseases, such as polio and measles. When an inactivated microbe is injected into the body, a limited number of specific antibody-producing lymphocytes are formed. Later, if infection occurs, the cells can immediately react, producing large amounts of protective antibodies quickly.

Skeletal System

The bones of early humans are in many cases all that is left for us to study. After centuries or longer, maggots and other decomposers remove all traces of living tissue, leaving only the mineral-like bone. Perhaps for that reason, we tend to think of the skeletal system as a nonliving structure only. However, bone in a living person is more than that. Bone is made up of microscopic living cells embedded in an abundant, hard, nonliving material. The living bone cells continually form new bone and recycle old bone. Bone both supports and protects soft body tissues, surrounding organs of the nervous, respiratory, digestive, and reproductive systems. It also helps regulate the levels of calcium and phosphate in the blood and lymph. Blood vessels that course through bone carry vitamins, hormones, and other substances that affect bone synthesis and breakdown.

Bone Cells

Most of the hard, nonliving substance of bone is a crystalline calcium phosphate compound called hydroxyapatite, $Ca_5(PO_4)_3OH$. About 99% of the body's calcium is in this form. A protein, collagen, makes up the remainder of bone. Fibers of collagen weave through the hydroxyapatite, giving the durable skeleton its flexibility. (Although durable and strong, bones are considerably elastic, helping the skeleton withstand impacts.)

Mature bone cells are of two types. Osteoblasts, the builders, synthesize new intercellular material and deposit it on existing bone surfaces. Osteoclasts break bone down and make its constituents available for reuse. These are large, multinucleated cells that actually glide through the pores in bone, stopping to secrete bone-dissolving acids and releasing calcium ions (Ca^{2+}) into the blood.

Bone Density

Before the age of about 35, bone synthesis predominates, but later on the breakdown by osteoclasts becomes greater and increases for the remainder of a person's life. The bone-dissolving activity of the osteoclasts creates cavities that in younger people are replaced by the depositing activity of osteoblasts. But as hormone levels decline with age, the synthesis can't keep up.

The extreme result of this imbalance is osteoporosis, a condition in which the bone loses so much density that it becomes porous and brittle. Genes, gender, diet, hormones, age, and activity all contribute to determining the severity of this condition.

Osteoporosis is most typical in postmenopausal women, who have low blood levels of the hormones estrogen and progesterone. Men can get osteoporosis, too, but their testosterone supports building greater bone mass. Though their risk is less than women's, men can get osteoporosis as their testosterone levels decline with age.

Several hormones and other drugs are successful in slowing osteoporosis, but exercise and a high intake of calcium can make it less likely to occur. The teenage years, marked by rapid body growth, are critical for producing bone mass. Stimulated by thyroid hormones, the osteoblasts use all of the calcium and protein provided by the osteoclasts, plus much of that in the diet. Dietary calcium is particularly valuable in the teenage years, during which 30% of the adult bone mass forms.

Muscular System

In recent years, Americans have become more muscle-conscious, with many of us working out in gyms, running, and swimming to build up muscles as well as to derive general health benefits.

Even for people with no interest in body-building, it is important to use the muscles daily. Inactive people can develop stiff muscles, setting up a cycle of even less activity, weight gain, and other health problems.

Types of Muscle Cells

All muscle cells are specialized for contraction. Those in striated muscle tissue make up the skeletal muscles, which are attached to bones and make voluntary movements possible. Those in smooth muscle tissue are found in the muscular walls of internal organs that contract involuntarily, such as the uterus, blood vessels, and organs of the digestive system. The cardiac (heart) muscle is made up of a special type of striated muscle; though it is striated, its contractions are involuntary.

Striated muscles are usually attached to tough connective tissue called tendons, which are attached to bones. Most muscles are in opposing pairs, with one of the pair contracting while the other relaxes. For instance, if you reach out toward something, the triceps muscle on the back of your upper arm contracts. If you then lift your hand to touch your shoulder, the biceps muscle on the front of your upper arm contracts, while the triceps relaxes. Both muscles are attached by tendons to bones in the forearm, and the bones' movements are controlled by those muscles.

Smooth muscle cells are spindle-shaped, have one nucleus, and contain the proteins actin and myosin. These molecules are arranged irregularly. Many of the smooth muscle cells are in two layers of tissue, with an inner circular sheet surrounded by an outer longitudinal sheet. The two layers work together in alternation to squeeze an organ's contents and move them along—as needed during digestion or childbirth, for instance.

The muscle cells in striated muscles show a regular pattern of striations when a **cross-section** is examined microscopically. Each cell is a fiber; thousands of fibers, bound by connective tissue, make up an entire skeletal muscle.

Sliding Filaments

Muscle fibers themselves are made up of thick (myosin) and thin (actin) filaments that are arranged in parallel and slide past each other. The striated pattern depends on the alternation of thick and thin filaments. The two types of filaments are connected by smaller cross-bridges of myosin. Each cell also contains several nuclei, which are pushed to the outside of the cytoplasm by the filaments.

The sliding filament theory explains the contraction of skeletal muscles. According to the theory, when a muscle contracts the filaments slide past each other, and the cross-bridges are rearranged. When the filaments slide apart again, the muscle relaxes.

Actin and myosin filaments slide past each other in smooth muscle cells also, but the fibrils are interwoven in a mass of fibers, not arranged in parallel. The actin and myosin filaments slide back and forth, but contraction is slower than in skeletal muscle and continues for a longer time.

Cardiac muscle appears striated, but the cells are shorter than those in skeletal muscle and have only one nucleus per cell. The heartbeat that continues from before birth to death is made possible by the involuntary contractions of cardiac muscle resulting from the movements of actin and myosin.

The Nerve–Muscle Connection

When a nerve impulse reaches a muscle, calcium is released at the neuromuscular junction. The calcium enters the muscle cells and activates the myosin bridges. Using energy from ATP molecules (see Chapter 1), the muscle filaments slide together, and the muscle contracts. Each fiber twitches (contracts) individually, with the contraction of the entire muscle being made up of individual twitches of many fibers. The twitch may be as fast as 7.5 milliseconds in skeletal muscle or as slow as 100 milliseconds in a smooth muscle.

Skeletal muscles and smooth muscles both respond to signals from nerves; smooth muscles may also contract in response to hormones. Adrenaline in the blood, for instance, can cause the smooth muscles in arteries to contract, narrowing the diameter of the vessels. Other hormones can aid in digestion, respiration, and other functions by causing various smooth muscles to contract.

Like smooth muscles, cardiac muscle responds to signals from both nerves and hormones, but the strength and rate of the heartbeat are controlled by the heart's **pacemaker** tissue.

Energy for Contraction

The contractions of muscles are fueled by the energy stored in ATP molecules during cellular respiration. Muscles also depend on a supply of oxygen and food and give off waste products that must be removed. All of these needs are supplied by the circulating blood. The color of a muscle indicates whether it has a good blood supply. Dark red beef, for instance, is well supplied, but the white muscle of a chicken breast is not.

Muscle fatigue results when muscle cells continue working too long. Too little oxygen is being delivered to the muscle for ordinary respiration to continue. Instead, the cells use **fermentation** to make ATP, and the end product (lactic acid) accumulates in the cells. This lowers the pH and reduces the muscles' ability to contract. Muscle fatigue is reversible; after the exercise stops, fermentation is no longer needed, and the lactic acid is broken down.

Muscle fatigue is common in anyone who is out of condition, but less so in those who exercise regularly. Exercise increases muscle size and benefits the circulatory system, which makes more nutrients available to the muscle cells and postpones the need for fermentation during exercise. In contrast, lack of exercise—such as that enforced curing prolonged bed rest—allows muscles to become shrunken, or atrophied.

The still-mysterious disorder called fibromyalgia is characterized by general, chronic pain in the muscles and skeleton. The pain and disability associated with the disorder causes patients to refrain more and more from physical activity, causing their muscles to become increasingly deconditioned. Treatment includes stretching and mild exercise to recondition muscles and reduce pain.[6]

SENDING CHEMICAL AND ELECTRICAL SIGNALS IN THE BODY

Distant parts of the body communicate with each other in two ways—chemical and electrical. The chemical secretions from glands act relatively slowly, the electrical and chemical signals from nerve cells very quickly. Each method has advantages for the organism, and glands and nerves work together to keep the body functioning and in chemical balance.

Many glands secrete their products through ducts that pass directly to their **target cells.** The eyes' tear glands and the skin's sweat glands are examples of these glands, called exocrine glands.

The endocrine glands, in contrast, secrete hormones—chemical messengers that act at a distance from their sources—into the bloodstream, which carries them to their destinations. Neurosecretory cells are neurons (nerve cells) that both carry electrical signals and secrete hormones into the bloodstream.

Though cells' possible functions are determined by their DNA, their actual functions occur partly in response to their environment. Cell functions are regulated by the neuroendocrine system (the nervous and endocrine systems). Neural and endocrine tissues work together to control and coordinate cell growth, activities, and division.

Body Chemistry

When the body's chemical reactions are in balance, every chemical is maintained within its normal range; for example, the normal range for potassium in the blood is 15–20 mg per 100 ml blood. Levels of a chemical may continually rise and fall in the blood and other organs, but the result is the equilibrium condition called homeostasis.

Most of the body's chemical processes affect a variety of substances through interactions with other processes. When the level of phosphorus in the blood builds up above its normal range, for instance, it combines with calcium to form calcium phosphate. That may lower the blood level of calcium so far that too little calcium enters the skeleton, which needs it for making new bone. If this condition continues, as it may when a patient has kidney failure and can't excrete phosphorus, osteoporosis may result. Some other conditions may lead to too much calcium in the blood (hypercalcemia), a cause of kidney stones and irregular heartbeat.

Stress can affect body chemistry as well as emotional responses. During the well-known "fight or flight" reaction to threat, the adrenal glands respond by producing the hormone adrenaline. Adrenaline in turn causes the breakdown of glycogen (the form of glucose that is stored in the liver) to glucose, which can be used for producing ATP in muscle cells. Energized by the ATP, the arm and leg muscles can contract to strike an enemy or beat a hasty retreat. Adrenaline also affects the digestive system, inhibiting the digestion and absorption that can await a quieter time.

Mental illness in many cases results not from purely emotional causes but from homeostasis gone awry. If a **neurotransmitter** substance is produced in too great or too small an amount, the effect on the rest of the nervous system may bring about the symptoms of clinical depression, **bipolar disorder,** or other mental illness.

Nervous and Endocrine Cooperation

The hypothalamus is part of the brain, but it secretes hormones that are stored in the pituitary. The hypothalamus and pituitary work together, helping coordinate actions of the nervous and endocrine systems.

The brain responds to some environmental signals that arrive through changes in hormones in the blood. The brain has receptors for thyroid hormone and six classes of steroid hormones, including estrogens and androgens.

The brain may also act as an endocrine gland. Growth of the body is regulated by the pituitary gland, as are many metabolic reactions, milk production, and other processes.

We all have internal clocks governing the times we feel sleepy, wake up without alarm clocks, are hungry, and so on. Because they are on a roughly 24-hour cycle, the clocks are called circadian. The clocks vary somewhat in individuals and can be reset by exposure to new patterns of light and darkness. Blind people, who have no external visual cues, may suffer from insomnia for this reason. A recent study of blind people showed that the hormone melatonin—produced within the brain by the **pineal** gland—can reset those clocks. When given either a placebo or melatonin, only the subjects who received melatonin reset their clocks to a normal cycle. At one time melatonin was expected to be a miracle cure for jet lag, but that has not proved to be the case. With the new results in blind people, scientists have new hope of finding melatonin's exact role in circadian cycles.[7]

Like melatonin, many hormones alter brain functions, so that the brain adjusts its own performance and control of the body's behavior to respond to a changing environment. For instance, rises in testosterone in males and estrogen in females influence the hypothalamus and pituitary to decrease the release of FSH and LH1 (discussed later). At the same time, changes in cell structure and chemistry occur, which lead to increased capacity to engage in sexual behavior.

Endocrine System

Unlike most body systems, the endocrine system is scattered throughout the body. The glands are defined as a system because they act in the same way and affect each other.

Hormones control the events of the menstrual cycle and many other body processes, such as changes in blood chemistry and responses to external stimuli. The hormone insulin, for example, is necessary for the cells to absorb sugar normally for use in cell respiration. In diabetic patients, too little insulin is made by the body, so sugar fails to enter the cells; it "backs up" into the blood and then spills into the urine.

Insulin

The islets of Langerhans are small clusters of endocrine cells embedded in the pancreas, which is both an exocrine gland and an endocrine gland. They secrete insulin into the blood, making it possible for the hormone to travel to any cell in the body.

Like all hormones, insulin must fit into receptor sites on a cell's surface or inside the cell before it can act, much as a key must exactly fit a lock before a door can be opened. Just as adenine pairs with only thymine or guanine in DNA, each hormone has a specific receptor molecule on or in the corresponding target cells. When the hormone and receptor have formed a combination, that event may result in the emission of a second messenger chemical that brings about the hormone's final effects in a cell. The hormone adrenaline, for instance, binds to its receptor on a cell surface, releasing a second messenger and initiating a series of chemical reactions within the cell that end with discharging a large amount of glucose into the blood.

Feedback mechanisms in the endocrine system act to make the levels of body chemicals rise or fall. Mechanisms that keep the body's chemicals in balance are called homeostatic.

Insulin, for instance, is secreted by the islets of Langerhans when blood levels of glucose are high, such as after you eat a doughnut or drink a soft drink. The sugars in foods are absorbed from the intestine and carried by the blood as glucose. Muscle, liver, and other cells targeted by insulin then take up glucose from the blood, lowering its level there. That causes the secretion of insulin to diminish as well.

Other endocrine cells in the pancreas secrete the hormone glucagon when the blood glucose level is low. Glucagon causes **glycogen** in the muscles and liver to be converted to glucose, raising the blood glucose level. By feedback, the pancreas knows it should then secrete less glucagon.

So insulin and glucagon act in tandem, influencing each other and keeping the blood glucose level within a normal range. That homeostasis is disrupted in diabetes or other conditions that affect the pancreatic endocrine glands. Three other hormones also affect blood glucose levels—growth hormone from the pituitary, adrenaline (epinephrine) from the **adrenal medulla,** and corticotropic hormones from the **adrenal cortex.**

Other Hormones

The pituitary is so important that it is sometimes called the master gland of the body. It is controlled in turn by the hypothalamus area of the brain. The pituitary gland's anterior lobe, the hypothalamus, and other tissues control the production of many hormones from other glands. Hormones that affect other hormones—TSH and ACTH are examples—are called tropic.

Figure 4.3
Human Endocrine System.

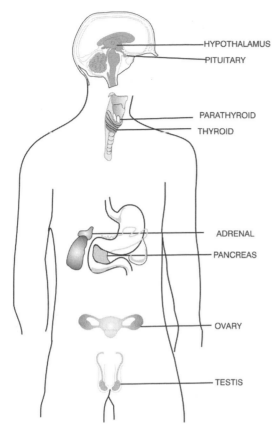

Vertebrates secrete a variety of hormones, as shown in Table 4.4. Although these hormones are associated with specific glands, a group of fatty acids called prostaglandins are made by most tissues of the body. These hormones affect the immune response, smooth-muscle contraction, and aggregation of blood **platelets.**

Nervous System

During the 1990s, thousands of cattle were sacrificed in Great Britain in an effort to control mad cow disease. Caused by infective proteins called prions, mad cow disease (bovine spongiform encephalopathy) causes damage to an infected **rumi-**nant's brain, including the loss of tissue that changes a firm brain into a spongelike mass. The animal may be uncoordinated and have difficulty standing, become aggressive, give less milk, and lose weight; usually it dies. The disease is spread primarily by cattle feed containing meat and bone meal from sheep infected with **scrapie.**

A similar condition called variant Creutzfeldt-Jakob disease (vCJD), which leads to **dementia,** sometimes attacks humans. Some victims in Europe may have contracted vCJD by eating infected beef that was contaminated with nervous system tissue.[8] The possible link is still under investigation.

Why do these diseases of the brain have such serious consequences for cattle and humans? It is because the brain controls all sensations, most movements and activities, and all thought and language. Without an intact brain, the body may continue to live for a time, but only in a vegetative state. Without a brain, a human being is not really human.

The brain and **spinal cord** together make up the central nervous system. The central nervous system is protected by the skull and vertebral column (spine). Pairs of cranial and spinal nerves enter the brain and spinal cord through openings between the bones. One nerve of each pair leads from a sensory receptor to the central nervous system and the other leads from it to an effector.

The peripheral nervous system consists of the nerves (bundles of neurons, or nerve cells) that extend outward from the brain and spinal cord to various parts of the body. Motor neurons of the somatic nervous system innervate the skeletal muscles, which the animal moves voluntarily. The cardiac and smooth muscles, which are generally moved involuntarily, are controlled by the

Table 4.4
Some Hormones of the Endocrine System.

Hormone	Endocrine Gland	Main Actions
Growth hormone	Pituitary (anterior lobe)	Stimulates bone growth and breakdown of fats; increases blood glucose
Prolactin	Pituitary (anterior lobe)	Stimulates milk production and secretion
Thyroid-stimulating hormone (TSH, or thyrotropin)	Pituitary (anterior lobe)	Stimulates thyroid
Adrenocorticotropic hormone (ACTH)	Pituitary (anterior lobe)	Stimulates adrenal cortex
Follicle-stimulating hormone (FSH)	Pituitary (anterior lobe)	Stimulates ovarian follicle in females, leads to spermatogenesis in males
Luteinizing hormone (LH)	Pituitary (anterior lobe)	Stimulates ovulation and formation of corpus luteum in females, interstitial cells in males
Oxytocin	Hypothalamus (hormones stored in and released from posterior pituitary)	Stimulates mammary glands and uterine contractions
Antidiuretic hormone (ADH, vasopressin)	Hypothalamus (hormones stored in and released from posterior pituitary)	Controls excretion of water
Thyroxin	Thyroid	Stimulates and continues metabolic activities
Calcitonin	Thyroid	Inhibits bones' release of calcium
Parathyroid hormone	Parathyroid	Stimulates bones' release of calcium, activates vitamin D, enhances absorption of calcium from intestine
Cortisol and other glucocorticoids	Adrenal cortex	Affect metabolism of all nutrients
Aldosterone	Adrenal cortex	Affects salt–water balance
Adrenaline and noradrenaline	Adrenal medulla	Increase rate and strength of heartbeat, dilate or constrict certain blood vessels, increase blood sugar
Insulin	Pancreas (islets of Langerhans)	Increases glycogen storage, lowers blood sugar
Glucagon	Pancreas	Stimulates breakdown of glycogen to glucose
Melatonin	Pineal	Affects circadian rhythms
Estrogens	Ovary, follicle	Develop and maintain sex characteristics in females, promote buildup of uterine lining
Progesterone	Ovary, corpus luteum	Promotes continued buildup of uterine lining
Testosterone	Testis	Develops and maintains sex characteristics in males, promotes spermatogenesis

sympathetic and parasympathetic divisions of the autonomic nervous system. The sympathetic division brings about effects needed in fight or flight responses; the parasympathetic brings about effects needed at other times, such as during digestion and reproduction.

The sense organs—eyes, ears, nose, tongue—contain specialized nerve cells that can detect light, sound, and specific chemicals, enabling animals to monitor what is going on in their environment.

Neurons

Each neuron consists of three parts—dendrite(s), cell body, and axon—that vary in size and shape depending on their role in the nervous system. Within each cell, **nerve impulses** always pass from the dendrite to the cell body to the axon.

A neuron's cell body is the central portion, which includes the nucleus. The dendrites are usually short and appear like offshoots of the cell body. In neurons that begin at sensory receptors, such as those for olfaction (smelling) in the nose or those for touch in the fingertips, dendrites may have a great many branches. They have fewer branches in neurons in other parts of the nervous system or in neurons leading to motor effectors. Effectors include cells in such organs as glands and muscles that act in response to stimuli from nerve cells.

Axons may be as long as 91 cm (3 feet) in, for example, the long nerve that extends from the lower spinal cord to the toes. Like dendrites, axons vary according to their role. If they are merely passing along a signal in a pathway from a sensory receptor to the spinal cord or to the brain, they may be simply shaped. If they are bringing about a response in some organ, they may be large and have many branches at the end, but there is never more than one axon per neu-

**Figure 4.4
Neuron.**

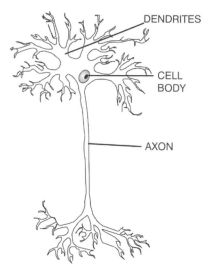

ron. It is usually much longer than the dendrites.

Surrounding the neurons are the glial cells, which have cell bodies and branched extensions. About 10 times as numerous as neurons, the glia are still under investigation. They were once thought to be merely for support and nourishment of the nerve cells, but it now seems that they play an active role in helping neurons communicate: Specialized glial cells called astrocytes signal the neurons to build and maintain synapses, the spaces between adjacent neurons.

Depending on the direction of the nerve impulse through the cell, a neuron is classified as a sensory neuron, motor neuron, or interneuron. A sensory neuron carries impulses from receptors toward the brain, spinal cord, or a ganglion. (A ganglion is a cluster of neuron cell bodies that coordinate incoming and outgoing impulses. Ganglia are especially important in invertebrate animals that have small or no brains.) Motor neurons carry impulses from those sources to effectors. Interneurons make the necessary connections among sensory neurons, motor neurons, and interneurons.

Like the hundreds or thousands of neurons of which they are composed, nerves can be sensory or motor. Some contain both types of neurons and are called mixed nerves.

Severed nerves cannot be restored; until recently it was believed that no new neurons could appear in adults. However, there have been several recent reports of partly differentiated rodent neurons being induced to revert to stem cells. These can potentially become new differentiated neurons.

Nerve Impulses

A nerve impulse begins with release of a chemical into the synapse between the axon of one neuron and the dendrites of another. Often the chemical is acetylcholine, a **neurotransmitter.** The acetylcholine combines with receptor molecules in the neuron's dendrites or cell body, which causes the cell membrane to become more **permeable** to sodium ions. During the resting state, when no impulse is passing through the neuron, the interior of the cell has a low proportion of sodium ions (+) and a high proportion of chloride (–) and potassium (+) ions compared with the exterior. As a result, the outside of the cell has a negative electrical charge relative to the inside. The ion concentrations are maintained by a mechanism in the membrane called the sodium pump, which counteracts the tendency of the ions to diffuse through the membrane. Acetylcholine changes that condition at the point of stimulation; the sodium ions can flow into the cell through ion channels (water-filled tunnels) that open up in the membrane, making the inside temporarily positive relative to the outside. (This change is called an action potential.) The permeability of the adjacent area is then changed, and the ions there flow through the membrane. Each area's permeability is altered in turn in a chain reaction, resulting in the passage of the nerve impulse all along the neuron. The change in permeability is very brief—less than a millisecond—and is followed by a temporary outward flow of potassium ions and then by the sodium pump's reestablishing the cell's resting state.

When the nerve impulse reaches the end of the axon, the cell secretes a neurotransmitter substance into the synapse with another cell or cells, and the impulse continues on its way.

Acetylcholine stimulates the transmission of an impulse, and so do substances such as dopamine, serotonin, and noradrenaline. Other substances may instead inhibit transmission (by causing the second cell's membrane to take in chloride ions and become more negatively charged in the interior). Any one neuron may receive signals from the axons of many other neurons—some signals that are stimulating, others that are inhibitory. The majority rules: If more of the signals lead to transmission than to inhibition, the impulse will begin. Otherwise, it will not.

Transmission of nerve impulses is fastest in axons that are surrounded by myelin, a substance made by specialized glial cells. The sheath of myelin is interrupted at intervals by openings (nodes), and the ions can move in and out of the neuron only there. So the impulse jumps from node to node, greatly increasing the velocity.

The popular antidepressant fluoxetine acts to prevent serotonin being taken back into the nerve cell that generated it, with the result that the message the chemical carries is transmitted for a longer than usual time. Recent research by Jessica Malberg and others at Yale University showed that rats treated with fluoxetine actually grew new, undifferentiated cells that became neurons. That suggests that fluoxetine and other drugs affecting neurotransmitters may someday prove useful in diseases like Alzheimer's.[9]

A nerve impulse travels along a circuit called a nerve pathway. These pathways may be very complex and link several effectors and receptors. The basic plan of a nerve pathway can be seen in a reflex arc, which is made up of a receptor, a sensory neuron, one or more interneurons in the central nervous system, a motor neuron, and an effector.

Because the reflex arc is simple, a nerve impulse can travel through it quickly, allowing a fast response to a stimulus. For example, if you touch a hot cooking pan or a flame, heat receptors in the skin immediately pass the nerve impulse through a sensory neuron toward the spinal cord. In the spinal cord, axon branches from the sensory neuron make synaptic connections with interneurons in the cord's interior (the gray matter). These quickly pass the impulse on to a motor neuron that ends in the hand's muscle fibers. The muscles contract, and the hand is reflexively pulled away from the heat. Normally all of this takes place in the fraction of a second, before the skin can be burned and even before you feel pain. At the same time, additional interneurons in the spinal cord pick up the impulse from the sensory neuron and pass it up the cord to the brain. The brain processes the information, recognizes pain and heat, and records the accident in memory.

Drugs usually affect the nervous system by changing the events at synapses. For example, cocaine blocks the reuptake of the neurotransmitters dopamine, noradren-a-line, and serotonin, causing them to remain in the synapses longer. Methylene-dioxymethamphetamine (MDMA, better known as Ecstasy) apparently releases sero-tonin and blocks its reuptake at synapses but decreases the serotonin and dopamine in the brain. Many antidepressants inhibit the enzymes that break down neurotransmitters. Alcohol increases the turnover of nor-adrenaline and dopamine, decreases transmission of impulses across synapses containing acetylcholine, increases transmission at synapses containing **GABA,** and increases the production of **beta-endorphin** in the hypothalamus. Nicotine has a variety of effects, depending on the dose and characteristics of the person using it. It attaches to receptors on the neurons' surfaces, and the actions vary according to the type of receptor and where it is located.

The Brain

In invertebrates, there is no central brain: Numerous ganglia or **nerve nets** coordinate sensory and motor impulses. In ancient vertebrates—fish, amphibians, reptiles, and birds—the brain consists of three main sections: the forebrain, midbrain, and hindbrain, which are approximately the same size. During the evolution of vertebrates, the hindbrain enlarged somewhat, and the forebrain became much greater in size. The major part of the forebrain, the cerebrum, has expanded greatly in primates, and in the human brain it is so large that little else is visible.

THE CEREBRUM

The cerebrum has several areas (lobes) devoted to memory, learning, coordination of speech, vision, hearing, taste, and so on. Nerve **tracts** from the spinal cord or from **cranial nerves** lead to the cerebrum through other parts of the brain. The cerebral lobes are named for the corresponding parts of the overlying skull.

One of the most interesting parts of the human cerebrum is Broca's area in the left temporal lobe, where speech is produced. It is active only on the left side of the brain. According to American linguist Noam Chomsky, author of *Language and Mind,* we construct language with nouns, verbs, and other parts of speech because our complex brains have an **innate** ability to use language, and this distinguishes humans from all other animals.

Various parts of the cerebrum are sensitive to stimuli received by the sense organs and transmitted to the cerebrum by cranial nerves. When the nerve impulses reach the cerebrum, they are passed to areas of the cerebral cortex, such as the visual cortex, auditory cortex, and so on. They are then translated by the brain into sights, sounds, odors, and other perceptions of the environment.

The cerebral cortex is a thin layer of gray matter overlying the cerebrum. (Gray matter is actually gray; it contains cell bodies of neurons. The white matter consists mainly of myelinated nerve tracts.) Though the gray matter in the cerebrum is on the surface, in the spinal cord the gray matter is in the interior, with the white matter on the surface.

The sensory cortex lies mainly posterior to the great **central sulcus** that crosses the brain from left to right, and the motor cortex lies mainly anterior to it. Each point on the sensory cortex is linked remotely to some sensory receptor in the body; each point on the motor cortex, to some motor effector in the body.

The cerebrum's left and right hemispheres—the left brain and the right brain—are nearly mirror images. In most people speech is controlled by the left brain, and spatial and musical ability by the right brain. They are connected by a thick bridge of nerve tracts called the corpus callosum. When the corpus callosum is severed, the two hemispheres can act independently.

Figure 4.5
Human Brain.

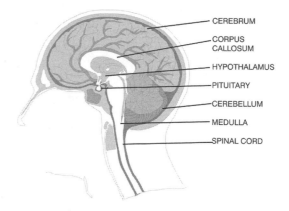

CEREBRUM
CORPUS CALLOSUM
HYPOTHALAMUS
PITUITARY
CEREBELLUM
MEDULLA
SPINAL CORD

The cerebrum is also where most learning and memory take place. Memories are apparently recorded when there are chemical changes at the synapses. Short-term learning leads to long-term memory that is established as nerve impulses pass through a circuit of cerebral structures including the **hippocampus, amygdala,** and **prefrontal cortex.**

Though the cerebrum is wired to perform many separate tasks and can handle many at once, there is occasional interference. Dr. Jan de Fockert and co-workers at University College London tested the ability of volunteers to ignore visual distractions, such as pictures of famous people, while trying to remember a series of numbers. When they were asked to remember more information—a strain on working memory—the volunteers found it harder to ignore the distractions.

The researchers interpreted their results to mean that when the brain must deal with a variety of distractions, such as those found during driving, it's important not to occupy too much available working memory. Using a cellular phone for a demanding conversation, for instance, might make it too difficult to ignore a distracting view or signs on stores while responding to important visual information, such as traffic lights and pedestrians entering crosswalks.[10]

Neurosurgeon Wilder Penfield (1891–1976) discovered much of what is known about the sensory and motor areas of the cortex by stimulating various points on his patients' cortexes during neurosurgery and recording their reports of the sensations and memories that were summoned forth.

It was evident at once that these were not dreams. They were electrical activations of the sequential record of consciousness, a record that had been laid down during the patient's earlier experience. The patient "re-lived" all that he had been aware of in that earlier period of time as in a moving-picture "flashback." . . .

A mother told me she was suddenly aware, as my electrode touched the cortex, of being in her kitchen listening to the voice of her little boy who was playing outside in the yard. She was aware of the neighborhood noises, such as passing motor cars, that might mean danger to him. . . .

D.F. could hear instruments playing a melody. I restimulated the same point thirty times. . . . Each time I restimulated, she heard the melody again. It began at the same place and went on from chorus to verse.[11]

OTHER PARTS OF THE BRAIN

In many animals, the cerebellum and cerebrum are about equal in size. In humans the cerebrum has become enlarged and covers much of the rest of the brain, but the cerebellum is still fairly large. Voluntary and involuntary movements are coordinated here. Nerve tracts from the spinal cord contain sensory and motor nerves controlling skeletal muscles, cardiac muscle, glands, and smooth muscles.

The thalamus and pons are not visible in the human brain from most angles. They are mainly areas of nerve tracts, where nerve impulses are routed to the cerebrum and cerebellum. The medulla oblongata, the most posterior part of the brain, is the enlarged end of the spinal cord.

The medulla, pons, and **midbrain** together make up the brainstem. This area controls vital body activities, such as breathing and heartbeat.

Cranial Nerves

Twelve pairs of cranial nerves extend from the brain to sense organs and other parts of the head and neck.

Sense Organs

Impulses travel to the brain from sensory receptors throughout the body, and from sense organs—eyes, ears, nose, tongue—that are specially adapted for receiving environmental stimuli.

When light enters the eye, it is focused by the lens onto the retina at the back of the eyeball. Specialized receptors in the retina, **rods** and **cones,** transmit information about shapes, amount of light, and color to the brain through the optic nerve. The left and right optic nerves cross, at a point called the optic chiasma, beneath the brain. Then the left optic nerve, carrying information that is mainly from the left **visual field,** enters the right side of the brain, and the right optic nerve, carrying information that is mainly from the right visual field, enters the left side of the brain. The nerves pass to the occipital lobe of the cerebrum, ending at the visual cortex. Here the cortex interprets the nerve impulses, producing a mental image. More impulses pass from the visual cortex to other parts of the brain, adding to the brain's store of memories and linking the new visual information with other knowledge. Impulses may pass through interneurons, leading ultimately to muscles

Table 4.5
Cranial Nerves.

Cranial Nerve	Area Innervated	Brain Area Where Nerve Ends	Functions
Olfactory	nasal epithelium	olfactory lobes	carries information about odors perceived
Optic	retina of eye	visual cortex of cerebrum	carries information about light, color, and shapes
Oculomotor	eye muscles	midbrain	controls eye movement
Trochlear	one eye muscle	midbrain	controls movement of one eye muscle
Trigeminal	branches to skin, lips, external ear, tongue, jaw muscles	medulla oblongata	perceives touch, controls muscles for chewing
Auditory	cochlea and vestibule of inner ear	medulla oblongata	perceives sound, aids in maintaining equilibrium
Facial	skin of face, taste buds, scalp, tongue	medulla oblongata	perceives taste
Abducens	one eye muscle	medulla oblongata	controls movement of one eye muscle
Glossopharyngeal	tongue and pharynx	medulla oblongata	perceives taste, controls salivation
Vagus	pharynx, heart	medulla oblongata	perceives taste, controls heartbeat, respiratory movements, peristalsis
Spinal Accessory	pharynx, larynx, neck and shoulder muscles	medulla oblongata	controls movements of pharynx, larynx, neck and shoulder muscles
Hypoglossal	muscles controlling tongue	medulla oblongata	controls tongue movements

and glands, causing the person to behave in response to what is seen.

A structure in the ear called the cochlea contains hairs that vibrate when sound waves reach them. The vibrations are translated into nerve impulses that go to the brain through auditory cranial nerves.

You perceive odors when molecules of something diffuse to your nostrils and activate receptors in the mucous membrane lining the nasal passages. Olfactory nerves lead from the receptors to the olfactory areas of the brain. Some animals—sharks, for instance—have extremely large olfactory lobes, befitting their predatory mode of

life. Sharks can detect the slightest concentration of blood in the water—less than one in a million parts. Humans depend much less on the sense of smell, and our olfactory lobes are very small in comparison with the cerebrum.

The sense of taste arises in the receptors in the tongue's taste buds. The taste buds can detect four basic tastes—sweet, sour, bitter, and salty. Specific parts of the tongue contain four types of taste buds, each type communicating information about one of the four tastes to cranial nerves that relay the report to an area deep in the cerebrum. Complex tastes may be made up of combinations of one or more basic tastes. The flavors we detect, in contrast to tastes, can be very subtle; they come from the combination of taste and odor.

Unlike the other senses, the sense of touch is not mainly transmitted through cranial nerves. Sensory receptors in the skin transmit nerve impulses about pressure, pain, cold, hair movement, and heat through peripheral nerves to the spinal cord, from where they travel to the brain for interpretation and action.

Behavior

Just as sense organs allow organisms to sense what is going on around them, effector organs—mainly muscles and glands—allow them to respond to their environment. Those responses are called behavior.

Innate (or inherited) behavior is something that an organism does without any learning. In animals that evolved early in geologic time and have no real brains, much behavior is innate. For example, *Drosophila* may perform fairly elaborate dances before mating, but no thought goes into the dancing; males and females are programmed to behave in certain ways in response to certain stimuli. Butterflies travel thousands of miles between winter and summer habitats in response to innate impulses.

In organisms with brains, behavior can be learned from changes in the environment or from parents. The larger the animal's brain, and the more elaborate its nervous system, the more of its behavior is learned; humans have relatively few innate behaviors. We do have some **reflexes,** such as the lower-leg jerk that follows when the kneecap is tapped, owing to reflex arcs.

This discussion is limited to the anatomical and physiological aspects of the nervous system. The human mind and consciousness are based on the nervous system, but they are also greatly affected by emotional states, life experiences, and learning. They are sometimes part of psychological studies.

CREATING A NEW INDIVIDUAL

All living organisms have some means of reproducing themselves. The methods requiring only one parent are called asexual; those having two parents, sexual. Whether asexual or sexual, the system making it possible to produce a new generation is called the reproductive system.

Asexual Reproduction

The simplest type of asexual reproduction is fission, a splitting of one organism into two, each of which is identical to the parent organism. Fission is common in one-celled organisms and in the animals that evolved earliest, such as sponges. **Algae** and other simple plants also reproduce themselves in this way. Reproducing algae, in fact, may quickly cover a pond or even a lake. Because the chromosomes are replicated prior to each mitotic division, all the cells produced by asexual reproduction are identical except when a mutation occurs.

A variation on this method is budding, a process in which a new plant or animal is formed from cells of an old one and then separates from it to become independent. Some sea anemones, for example, form new individuals in this way.

Some organisms—sponges and jellyfish, for instance—practice both asexual and sexual reproduction. When they use vegetative reproduction (an asexual process), cells broken off from an old organism divide and differentiate to become new organisms. Or they can begin producing haploid gametes by meiosis and release the gametes into the water. These fuse, restoring the diploid number, and develop into a new sponge or jellyfish.

TOOLS AND TECHNIQUES: NEW IMAGES FOR MEDICAL DIAGNOSES

For nearly a century, X-rays have been important for use in diagnosing some disorders. An X-ray can show bones, and metals that have been injected or swallowed, but it is of little use in providing information about soft tissues. If you have examined an X-ray, you have probably noticed that except for bone and metal, much of the picture is muddy-looking. For showing the brain, especially, tomography—a technique for getting clear X-rays of deep internal structures by focusing on just one plane within the body—has been more useful.

In linear tomography, the X-ray tube is moved over the body, emitting radiation, and the film is also moved, but in the opposite direction. Most of the structures in the X-ray picture are blurred by that motion. Just one area, on a plane partway between the tube and the film, is in focus. As the tube and film are moved, different points along the plane come into focus; and so a series of pictures is produced, showing the various points.

A variation of that technique is called computerized axial tomography (CAT). During the early 1970s the CAT scan was developed by scientists in Great Britain and the United States. In the years since, it has become a common part of medical diagnosis. Unlike linear tomography, CAT scanning uses no film. Instead, the results of the X-ray are recorded as a pattern of electrical impulses by a radiation detector. The X-raying is repeated many times, with the pattern recorded each time. Data from thousands of points are integrated by a computer, which converts them into mathematical information about the density of tissues at all the points. Those densities are then shown as points of varying brightness on a television TV screen, giving a detailed cross-sectional image of the organ being examined.

In magnetic resonance imaging (MRI), no X-rays are needed. The patient lies down in a large tube that is actually a magnet. The strong magnetic field it generates causes cer-

tain atomic nuclei—mainly those in hydrogen atoms—to align with the field, much like compass needles align with the Earth's magnetic field. Because hydrogen is found in the greatest proportion in water molecules, soft tissues are affected more than bone and other dense tissues that contain little or no water. Radio waves are then used to stimulate the magnetized nuclei in the patient's tissues. As the tissues respond to the radio waves, a computer picks up signals from them and converts them to visible images.

Both chemical and structural information can be obtained with MRI, which is very helpful for examining the brain, spinal cord, pelvic organs, and urinary bladder. Athletes and others have been aided by MRI studies of their joints and tendons that provided information about the fluid and solid materials present.

MRI is a very safe procedure, but it is also very expensive. Also, many patients find it unpleasant because of having to lie without moving in a narrow tube throughout the long procedure. Because it generates a strong magnetic field, MRI cannot be used in patients having pacemakers or having any metal in nervous tissue (such as having a rod in the spine or a metal plate in the skull).

In spite of these drawbacks, MRI is now used often for diagnosis. It has nearly replaced arthrography (visualization of cartilage or ligaments that have been injected with dye) and myelography (visualization of the spinal cord following injection with dye). One form of MRI even can produce an image of a patient's blood as it flows through arteries and veins, and three-dimensional images can be created for some purposes.

Both CAT and MRI techniques yield two-dimensional pictures of cross-sections of the body, and so a radiologist must examine and interpret the images in sequence. However, three-dimensional holograms can now be produced by a computer from CAT or MRI digital data. Holograms are especially helpful to vascular surgeons and neurosurgeons who

CONTINUED

need to pinpoint the exact location of lesions before performing surgery.

A third important diagnostic scanning tool is positron emission tomography (PET). In this procedure, as in traditional X-rays, radioactivity is involved: A chemical compound is labeled with a short-lived, positron-emitting radioactive portion and injected into the patient's body. During radioactive decay of the compound, positrons are destroyed by electrons, producing gamma rays. The gamma rays are detected by radiation detectors that are set up on opposite sides of the patient, and a computer integrates the information to produce images of the scanned organs. PET scans are used for studying functions of the brain and heart.

Altogether, these techniques have transformed the field of radiology, producing images that are more informative and being safer for patients than traditional X-rays (Shoolery 2002).

Source: James Shoolery, interview by the author, Half Moon Bay, Calif., 2 April 2002.

Sexual Reproduction

Bacteria, as well as many plants and most animals, reproduce sexually. Biologists think sexual reproduction evolved and was selected for because of the genetic diversity it makes possible.

Meiosis, discussed in Chapter 3, is one stage in gametogenesis, or the formation of gametes. In females the process is called oogenesis (formation of ova); in males, spermatogenesis (formation of sperm). Immature cells in the **ovaries** and **testes** go through several stages to produce ova and sperm cells. Ova are large cells containing much cytoplasm; sperm cells are small, being specialized for swimming to and fertilizing an egg.

Fertilization of an ovum by a sperm always occurs in a watery medium. In aquatic animals and plants this can actually take place in a pond, for instance. In terrestrial animals, internal fertilization provides enough water for the sperm to swim to the egg. (Land animals needed to evolve a variety of structures and changes in physiology before they could leave their aqueous environment permanently. The first group to begin the transition, amphibians, continued to fertilize their eggs in the water, though their adult forms were terrestrial. Reptiles and other later groups use internal fertilization.)

The Human Reproductive System

In humans, fertilization usually takes place in one of the oviducts, or Fallopian tubes, where the egg travels from an ovary. Seminal fluid that has been released during the male's ejaculation flow into the vagina, through the cervix, into the uterus, and into the oviducts. (Because the vagina's muscular contractions help draw the seminal fluid into the reproductive tract, this can occur even if the penis is not inside the vagina.)

Sexual intercourse often ends in an orgasm for one or both partners. Orgasm includes involuntary muscle contractions and the release of seminal fluid by the man.

One of the sperm cells in the seminal fluid reaches the egg and fertilizes it, forming a fertilized egg that contains a nucleus with the $2n$ number (46) of chromosomes. The fertilized egg later implants in the lining of the uterus (the endometrium), where it will develop first into an **embryo** and later into a **fetus**. If there is no fertilization, the uterine lining is shed (menstruation).

The Menstrual Cycle

Though there is individual variation in fertility, a woman normally goes through a monthly cycle, controlled by hormones, in which about every 28 days the uterine lining

Figure 4.6
Human Female Reproductive System.

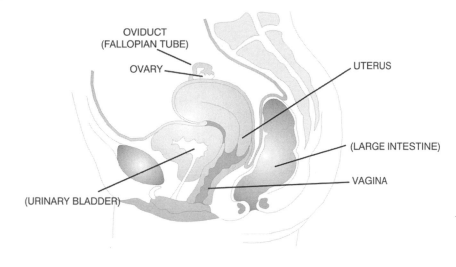

is built up as preparation for pregnancy. Usually just one egg is released per month. The hormone estrogen is at a low level as menstruation begins, as is progesterone; both hormones are produced by the ovaries. The low levels bring about **secretion** of gonadotropin-releasing hormone (GnRH) by the hypothalamus. GnRH travels from the hypothalamus to the **pituitary gland** through blood vessels and stimulates the release of the pituitary hormones follicle-stimulating hormone (FSH) and luteinizing hormone (LH1). These influence the ovary, and an egg starts to mature inside a **follicle** there.

The follicle is also stimulated by FSH and LH1 to release estrogen into the blood, traveling to the uterus and causing the uterine lining to thicken and become filled with blood vessels. After about 14 days, during which these hormonal effects continue, secretion of LH1 by the pituitary suddenly increases, and as a result the follicle bursts open and releases the egg—the event known as ovulation. The empty follicle becomes the yellow corpus luteum, which releases estrogen and progesterone. These

bring about continued growth of the uterine lining and cause the hypothalamus and pituitary to secrete less GnRH, FSH, and LH1.

Whether the egg is fertilized determines the next steps in the sequence. If it is fertilized and implants in the uterus, it begins to form a **placenta,** which makes human chorionic gonadotropin (HCG). The corpus luteum, under the influence of HCG, continues to release estrogen and progesterone, maintaining the uterine lining during pregnancy.

If the egg is not fertilized, the corpus luteum degenerates, causing a lowering of the levels of estrogen and progesterone. Because that decreases the supply of blood to the uterine lining, the lining breaks down and flows out of the body as menstrual fluid. Then another menstrual cycle begins.

The Male Reproductive System

Human males, in contrast to females, normally produce gametes continually after **puberty.** Health and environmental conditions may affect a male's fertility, however;

Figure 4.7
Human Male Reproductive System.

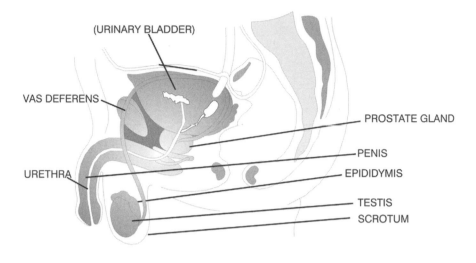

for instance, if sperm cells become too warm, they cannot fertilize an egg. The sperm cells are produced in the **seminiferous tubules** in the testes, the organs inside the **scrotal sac.** Because the sac extends outside the body, the sperm can remain cooler than the normal 37°C (98.6°F) temperature of the body.

From each testis the sperm cells go to the **epididymis,** where they are partly mobilized. This structure is continuous with the vas deferens, the tube leading to the **urethra.**

The human male's sex organs are much more obvious than the female's. (The sex organs of females—all internal—are the ovaries, oviducts, uterus, and vagina.) The penis, which serves as the external opening for both the urinary tract and the reproductive tract, is made of a spongy tissue that is rich in blood vessels. It becomes filled with blood during sexual excitation and empties after orgasm. During the ejaculation that accompanies orgasm, fluid from the **bulbourethral glands,** the **seminal vesicles,** and the prostate are mixed and flow out through the urethra as semen.

Several hormones control sperm production. LH1 stimulates the production of testosterone by cells in the testes. Testosterone and FSH bring about the production of sperm. Those and other hormones influencing sperm production are regulated by a feedback system. Hormones from the hypothalamus and pituitary gland interact in this system.

Secondary Sex Characteristics

Hormones also bring about male secondary sex characteristics, such as a low voice, enlarged larynx (Adam's apple), and chest hair. The collective name for male sex hormones is *androgens.*

In females, secondary sex characteristics include a relatively high voice, enlarged breasts, and wide pelvis. These, too, are controlled by hormones. Estrogen and progesterone, as well as FSH, LH1, and GnRH, affect females' appearance and physiology.

Table 4.6
Birth Control Methods.

Method	Effect	Effectiveness*
Complete abstinence	No sexual activity	100%
Hormonal capsule implantation under skin	Egg release prevented	99.9%
Tubal ligation	Prevents movement of oocyte to uterus	99.5–99.9%
Vasectomy	Prevents release of sperm	99.5–99.9%
Intrauterine device (IUD)	Prevents egg from being fertilized or implanted	97.4–99.2%
Hormone injections	Egg release prevented	99.7%
The Pill (estrogens and progesterone)	Prevents maturation of follicle and ovulation by inhibiting hormone secretions	95–99.9%
The "minipill" (progesterone only)	Prevents conception by changing cervix and uterus	90–97%
Withdrawal	Penis removed from vagina before ejaculation	81–96%
Cervical cap with spermicidal jelly	Prevents sperm from entering uterus; jelly kills sperm cells	Up to 90%
Emergency contraception (high dose of estrogen)	Probably prevents implantation	75–89%
Condom (male or female) or diaphragm (female) with spermicidal jelly	Prevents sperm from entering uterus; jelly kills sperm cells	79–98%
Contraceptive sponge containing spermicide	Prevents sperm from entering uterus; jelly kills sperm cells	80–87%
Vaginal foam (jelly alone)	Acts as barrier to sperm	70–80%
Condom (male) alone	Prevents sperm from entering vagina	64–97%
Rhythm (abstinence during expected time of ovulation)	Prevents pregnancy if time of ovulation is known	53–86%
Douche	Washes out sperm from vagina	15% or more

*Effectiveness is defined as the percent of women users who do not become pregnant each year.
Source: www.plannedparenthood.org/bc (Planned Parenthood Web site)

Birth Control

Birth control has been practiced with more or less success for centuries, but became much easier during the 1960s, when the Pill was introduced. The estrogen and **progesterone** inhibit the hormone secretions that normally lead to maturation of the follicle and to ovulation, and so ovulation itself does not occur. Today a smaller dosage of the hormones is in the Pill, and varied other birth control methods (hormonal and nonhormonal) are available as well.

NOTES

1. N. M. Brook et al., "The ParaHox Gene Cluster is an Evolutionary Sister of the Hox Gene Cluster." *Nature* 392 (1998): 920–922.

2. J. E. Brody, "Health Sleuths Assess Homocysteine as Culprit," *New York Times,* June 13, 2000.

3. I. H. Mokdad et al., "The Spread of the Obesity Epidemic in the United States, 1991–1998," *Journal of the American Medical Association* 282 (1999): 1525.

4. T. N. Robinson, "Television Viewing and Childhood Obesity," *Pediatric Clinics of North America* 48 (4) (2001): 1017–1025.

5. D. Barboza, "Rampant Obesity, a Debilitating Reality for the Urban Poor," *New York Times,* December 26, 2000, p. D5.

6. J. E. Brody, "Fibromyalgia: Real Illness, Real Answers," *New York Times,* August 1, 2000.

7. Associated Press, "Hormone Can Regulate Internal Clock," October 12, 2000.

8. Paul Brown et al., "Bovine Spongiform Encephalopathy and Variant Creutzfeldt-Jakob Disease: Background, Evolution, and Current Concerns," *Emerging Infectious Diseases* 7(1) (Jan–Feb 2001).

9. "Growing Hope," *Economist,* December 7, 2000.

10. Jan W. deFockert et al., "The Role of Working Memory in Visual Selective Attention," *Science* 291 (2001): 1803–6.

11. Wilder Penfield, *The Mystery of the Mind* (Princeton, N.J.: Princeton University Press, 1975), pp. 21–22.

5

A Survey of Organisms

Chapter 4 emphasized the systems in the human body, but now the perspective will be much broader. This chapter will survey the whole collection of organisms in the biosphere. Some of the organs and systems in other organisms are remarkably similar to those in humans, and some are completely different. However, the same evolutionary process underlies their characteristics: Each kind of organism has come to be what it is as the result of adapting to changing environments by solving specific problems.

In the long struggle for survival on Earth, living things have continually had to overcome new obstacles to survive or move into new surroundings. Earth itself has warmed and cooled; seas have risen and receded; mountains have been uplifted and eroded. Organisms had to change often enough to adapt to the new physical conditions and the changing organisms around them. Most of them, unable to make the necessary changes in changing environments, became extinct. Others became very different and survived. This chapter will review the groups of organisms that are currently successful.

Many classification systems and groupings are possible. The five-kingdom system used in this chapter is based on that in *Five Kingdoms: An Illustrated Guide to the Phyla of Life,* by Lynn Margulis and Karlene Schwartz (New York: Freeman, 1998).

THE BEGINNINGS OF CLASSIFICATION

Looking across a landscape, you may at first see a general pattern of shapes and colors. Soon, though, you will begin to identify the various organisms in the landscape—pine trees, seagulls, or prairie flowers, perhaps. You may think "least tern," "rose-breasted grosbeak," or "pileated woodpecker," not just "bird."

Since ancient times, humans have struggled to identify plants and animals, for generally selfish reasons. Plants have been used as sources of medicines as well as foods. First-century Greek herbalist Dioscorides (c. 40–90 A.D.), for instance, taught that mandrake roots could be used as an anesthetic before surgery was performed, but "used too much they make men

Figure 5.1
Geologic Time. As the Earth's environment has changed through time, organisms have adapted or died out.

Era	Period	Epoch	Earth History*	Time † (of beginning)
Cenozoic	Quaternary	Holocene	Human-built environments	11,000 (not in millions)
		Pleistocene	Ice Ages followed by warming	1.8 MYA
	Tertiary	Pliocene	Continental elevations and cooling	5 MYA
		Miocene	Moderate temperatures; grasslands and plains	23 MYA
		Oligocene	Mild temperatures; rain; erosion of mountains	38 MYA
		Eocene	Very warm temperatures; mountains built	54 MYA
		Paleocene	Asteroid; extinction of dinosaurs	65 MYA
Mesozoic	Cretaceous		Mild to cool temperatures; inland seas and swamps spread; Himalayas, Rockies, Andes built	146 MYA
	Jurassic		Cool, then mild temperatures; shallow seas and continents; Sierra Nevada mountains built	208 MYA
	Triassic		Cool and dry; continents elevated; many deserts	245 MYA
Paleozoic	Permian		Cold and dry; continents elevated; Appalachians built	286 MYA
	Carboniferous		Inland seas; swamps; warm and moist; coal deposits	360 MYA
	Devonian		Inland seas; mild; first forests	410 MYA
	Silurian		Mild; continental seas and reefs	440 MYA
	Ordovician		Warm; land submerged	505 MYA
	Cambrian		Mild; land submerged three times	544 MYA
Precambrian		Proterozoic	Variable climate; volcanoes; mountains built; glaciers formed	2500 MYA
		Archaean	Little oxygen	3800 MYA
Hadean			Formation of Earth	4500 MYA

† MYA = millions of years ago

* Based on C. P. Hickman et al. *Integrated Principles of Zoology.* St. Louis: Mosby, 1974.

speechless."[1] Anyone who wanted to collect certain kinds of plants needed to know how they differed from similar plants, especially if the similar plants were ineffective or poisonous. Even today, people are occasionally poisoned by eating mushrooms mistakenly identified as being safely edible varieties.

One of the first schemes for classifying living things was devised by Greek philosopher and scientist Aristotle (384–322 B.C.), who paid little attention to plants but described about 500 kinds of animals. He classified animals in a ladderlike *Scala Naturae* or Chain of Being that assumed humans were almost at the top of the ladder. The ladder's descending rungs were God, humans, mammals, oviparous with perfect eggs (such as birds), oviparous with non-perfect eggs (such as fish), insects, plants, and nonliving matter.

Fabled and Real Organisms

Animals were regarded with much superstition through the Middle Ages, and elaborate bestiaries from that time show an eclectic assortment of real and fabled animals. There was a famous ant-lion, the offspring of a lion and an ant, for instance. Because a lion cannot eat plants, and an ant cannot eat meat, the ant-lion had no food source and was doomed to starvation.

When Europeans began discovering other lands and the organisms living there, naturalists started looking more carefully at the real plants and animals around them, as well as at those in faraway places. What they learned was at first added to fables about unicorns, mermaids, and sirens, but gradually zoologists became realistic in naming and classifying animals.

Naturalists in different countries found it difficult to communicate about their discoveries because they spoke different languages. Even when they named plants and animals, they could not be sure that *la cucaracha* was the same animal as a cockroach; because they classified animals by alphabetical order, classification systems in different countries scarcely resembled each other. Some sort of international standards for naming and classifying were needed. In botany, some of those standards were provided by a young Swedish physician, Carl Linnaeus (1707–1778).

Taxonomy

To provide useful medicines, eighteenth-century physicians needed to know a great deal about plants and their properties. For most physicians, botany was simply a necessary tool for the practice of medicine, but Linnaeus was always more interested in botany itself than in medicine. Even as a child he was an avid collector of plants. As an impoverished medical student, he abandoned his studies for a year to organize a plant-collecting expedition to Lapland.

Eventually Linnaeus became a successful physician—so successful that later in his life the king of Sweden knighted him and dubbed him Carl von Linné. (He is also sometimes referred to as Carolus Linnaeus.) However, he managed to continue the work in botany for which he is famous today. Linnaeus, in fact, is often called the father of taxonomy. He devised a system for naming, ranking, and classifying organisms that is still in use, though the system has been modified by scientists since his time.

Linnaeus inspired a generation of botany students. According to a description of the field trips he led:

when he annualy botanized in summertime, he had hundreds of auditors who collected herbs and Insects, arranged observations, shot birds, made notes in the minutes, and when they from morning 7 o'clock to evening 9 o'clock . . . had Botanized, they returned to town with Flowers in their hats, and with drums and horns followed their leader to the Garden through the entire town.[2]

Nineteen of those entranced botany students went out on trade and exploration voyages to all parts of the world. It was the Age of Discovery, and they were able to take advantage of it for collecting purposes. One student, Daniel Solander, became the naturalist on Captain James Cook's first around-the-world voyage. Another, Anders Sparrman, was a botanist on Cook's second voyage. Most brought back or sent plants for their mentor's collections.

Both loving nature and being very religious, Linnaeus believed it was possible to understand God's wisdom by studying the living world. He wrote in the preface to a late edition of *Systema Naturae: "Creationis telluris est gloria Dei ex opere Naturae per Hominem solum*—The Earth's creation is the glory of God, as seen from the works of Nature by Man alone." For Linnaeus, it followed that the task of naturalists was to construct a "natural classification" that would demonstrate the divine order in the universe.

The Linnean System

Perhaps surprisingly, Linnaeus based his plant taxonomy entirely on plants' sexual organs. Biological classifications today are based instead on how organisms are related, with evidence from all parts of the organism and all stages of its development. We still use the Linnean system's **hierarchical** classification and **binomial** nomenclature, however.

Aristotle established the term *genus* and referred to the *differentio specifica* of each type of organism, but Linnaeus used the genus and species concepts as the bases of naming organisms and as the lowest levels of a hierarchical classification system.

He grouped similar species into a genus, similar genera into an order, and so on up to the kingdom level.

Binomial nomenclature refers to the method Linnaeus established for giving a unique name to each species. Earlier scientists had used binomial names (usually in Latin), but Linnaeus did so consistently according to genus and species. He named the wild briar rose, for instance, *Rosa canina. Rosa* is the genus of roses; *Rosa canina* is the wild briar rose species, and *Rosa damascena* is the damask rose species. When several species are mentioned together, or the genus name is obvious, a binomial may be given with only the first letter of the genus name—for example, *R. canina* and *R. damascena.* (Because the same **specific epithet** may be used for species in different genera, the words *canina* and *damascena* alone are insufficient.)

Today, Linnaeus's system of binomial nomenclature is so well established that we use it without thinking about it. We may refer casually to people as *Homo sapiens* or to dogs as *Canis familiaris.* Following Linnaeus, we italicize both names and capitalize only the first one (the genus). When a subspecies is established, we add a third name as a modifier. For instance, *Homo sapiens sapiens* is a modern human and *Homo sapiens neanderthalensis* is a Neanderthal—subspecies close enough to be in the same species, but with a slight genetic distance.

THE FIVE-KINGDOM SYSTEM

Linnaeus visualized organisms as being divided into just two broad groups, or kingdoms—plants and animals. For scientists and students today, classification is not that simple. Biologists have proposed many groupings, always with the goal of establishing broad groups having a common evo-

Figure 5.2
The Levels of Biologic Classification.

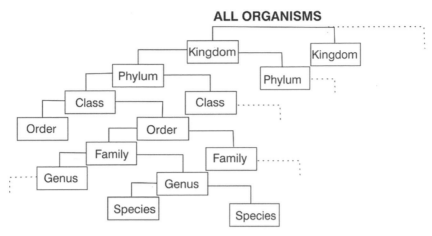

lutionary lineage. Because it isn't always possible to determine relationships exactly, different scientists have set up different groupings; as many as 19 different kingdoms of organisms have been suggested at various times.

Today organisms are classified into a hierarchy of groups and subgroups based on similarities that reflect their evolutionary relationships. When you look at plants, animals, or other living things, you see the organisms as they are at that moment. However, each one is like one frame of a film, the result of long years of evolution and the source of future forms. Because living and nonliving environments change over time, the organisms within them are molded by those changes.

Charles Darwin was awestruck by that vision of nature, ending *The Origin of Species* with the lines

It is interesting to contemplate a tangled bank, clothed with many plants of many kinds, with birds singing on the bushes, with various insects flitting about, and with worms crawling through the damp earth, and to reflect that these elabo- rately constructed forms, so different from each other, and dependent upon each other in so complex a manner, have all been produced by laws acting around us. . . . There is grandeur in this view of life . . . that, whilst this planet has gone cycling on according to the fixed laws of gravity, from so simple a beginning endless forms most beautiful and most wonderful have been, and are being, evolved.[3]

Within each of the kingdoms are different groups called phyla; within each phylum, different classes; within each class, different orders; within each order, different families; within each family, different genera; and within each genus, different species. Organisms in the same kingdom, such as tigers and snails, may be only distantly related or may be near kin. Organisms in the same species, such as dogs and wolves, are so closely related that they can interbreed. Species is the most fundamental unit of classification.

In learning about classification, it is important to remember that the kingdoms and other groups are an artificial constraint by humans on the underlying natural phe-

Figure 5.3
The five-kingdom system currently in use by many biologists.

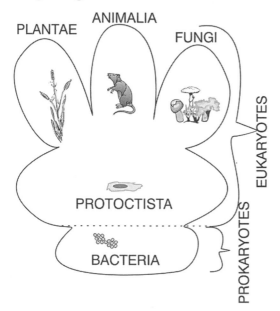

PLANTAE   ANIMALIA   FUNGI

PROTOCTISTA

BACTERIA

PROKARYOTES   EUKARYOTES

nomena. As new information accumulates, older classifications are revised and organisms are reclassified.

In spite of the tentative and artificial nature of classification, it is useful to refer to related groups of organisms in discussion and experimentation. In this book is a scheme that is used in many textbooks and that has a sound base in evolutionary theory. It is the five-kingdom system associated mainly with biologists R. H. Whittaker (1920–1980) and Lynn Margulis. The five kingdoms they proposed are Plantae, Animalia, Fungi, Protoctista, and Bacteria.

First Life on Earth

Even as late as the 1900s, some people failed to realize that life comes only from life. From commonsense observation they thought that vinegar or stagnant water could give rise to worms, for instance. Earlier than that, everyone knew that if meat was left in the open, maggots (fly larvae) appeared in it. They were thought to appear by **spontaneous generation.**

That idea, however, was disproved by Italian biologist Francesco Redi (1626–1697). He used controlled experiments to show that when cooked meat was left exposed to air but covered with fine gauze, no maggots appeared in the meat. Meat that was left uncovered was accessible to flies that laid eggs on it, giving rise to maggots.

Apparently maggots can come only from fly eggs, which are too small to be visible on the meat. The eggs had come from flies that came from eggs, and so on. Or, "all life from life."

But where did the first life on Earth come from? Did it arrive on meteorites from elsewhere in the universe (the panspermia hypothesis), as some scientists believe? Or did it come from a Frankensteinlike energizing of nonliving material? The answers to those questions are yet to come. We do, however, have some idea of what has happened to life since it first arose.

Prokaryotes

The Earth is about 5 billion years old, according to our best evidence. Life itself appeared about 3.5 billion years ago, probably in the form of bacterialike organisms. The oldest fossils found are filaments from rocks in Western Australia, and round fossils in southern Africa. Both types resemble bacteria.

The earliest organisms were prokaryotes—that is, they had no defined nuclei or other membrane-bound organelles in their cells. (Mitochondria and chloroplasts also are surrounded by membranes.) The first eukaryotes appeared about 1.2 billion years ago. The prokaryotes and eukaryotes are the two great divisions of all living things;

Table 5.1
The Kingdoms of Organisms.

Kingdom	Groups	Examples	Characteristics
Bacteria (prokaryotes)	Archaeans, bacteria	Extremophiles, *Streptococcus*	No membrane-bound organelles or nuclei; asexual reproduction of cells; most are unicellular; autotrophic or heterotrophic
Protoctista (first eukaryotes)	Algae, slime molds, flagellates, and others	*Amoeba*, green algae, *Euglena*	Reproduction of cells asexual or by mitosis; unicellular or multicellular; autotrophic or heterotrophic, great variation in all characteristics
Plantae	Mosses and liverworts, ferns, conifers, flowering plants, and others	*Acer* (maple), Boston fern, horsetails	Cells surrounded by cell walls; multicellular; food stored as starch; chloroplasts in cells; autotrophic; develop from an embryo
Fungi	Molds, lichens, yeasts, and others	*Rhizopus* (bread mold), baker's yeast, bracket fungus	Develop from spores; no undulipodia; unicellular or multicellular; heterotrophic
Animalia	Arthropods, chordates, worms, and others	Octopus, tarantula, *Homo sapiens*	Most are motile; multicellular; heterotrophic; develop from a blastula

all organisms today are in one group or the other.

Because they cannot live independently, viruses are not listed with the organisms in Table 5.1. However, viruses may become part of living cells by merging with a host's DNA. Among the viruses are the rhinoviruses that cause the common cold, and HIV, the cause of AIDS.

Bacteria

In some classification systems, this kingdom is called Monera. The bacteria fall into two broad groups, Archaebacteria and Eubacteria, based on the structure of their ribosomal RNA.

ARCHAEA

The Archaea were named by Carl Woese, a biologist at the University of Illinois who did some of the early studies of the organisms. He also made them one kingdom in his own three-kingdom classification system. (Woese's other two kingdoms are the prokaryotes and eukaryotes.)

Archaea are the oldest organisms on the earth, and their characteristics represent the adaptations needed for surviving extreme conditions about 3.8 billion years ago. During that time, the Archaean epoch, there was less than 1% of the amount of atmospheric oxygen in today's atmosphere available for the new living things to use. The archaeans also had to cope with terrific heat: Volcanoes poured forth molten lava that gradually cooled, forming the first terrestrial rocks.

Archaeans—sometimes called extremophiles—are like eubacteria in many ways, such as being one-celled and lacking cell nuclei. On the other hand, they appear to be more closely related to eukaryotes than to bacteria. They may be the oldest organisms on Earth.

Table 5.2
Major Groups of Bacteria.

Group	Examples	Characteristics
Archaebacteria	"Extremophiles," such as the halophiles and thermophiles	Adapted to extreme environmental conditions; unique biochemistry
Eubacteria	*Escherichia coli, Staphylococcus*	May be free-living or parasitic

Archaeans live in an impressive variety of harsh environments, such as undersea hyperthermal vents and the surface waters off Antarctica. Though the temperatures tolerated by most organisms range from 4°C to 50°C (39°F to 122°F), the range for extremophiles is –1.8°C to 113°C (29°F to 235°F).[4] Within this group, some can tolerate heat (thermophiles), others can live in very cold climates (psychrophiles), and still others can withstand high pressure (barophiles). Some of them are mesophiles, or organisms that live in what we would consider moderate conditions.

Most organisms can't live where the pH is lower than 5 or higher than 9. Extremophiles, though, can be found in very acidic or very basic environments. The acidophiles live in very acidic areas, such as volcanic pools and hot vents on the sea floor, where the pH may be less than 1. The alkaliphiles live in environments having a very high pH (more than 11). Halophiles can tolerate very salty environments, such as the salt lakes in the western United States.

Many extremophiles live in conditions that are unusual in more than one way. Thermoacidophiles withstand both the heat and the acidity of hot springs, and haloalkaliphiles live in soda lakes that are both salty and alkaline.

Before about 2 billion years ago, the concentration of oxygen in the atmosphere was less than 0.001. Some of the archaeans may have been cyanobacteria, which could photosynthesize and build up oxygen in the atmosphere. As a result, they paved the way for the emergence of the oxygen-using bacteria.

EUBACTERIA

Eubacteria may be free-living autotrophs or parasitic heterotrophs. We are more aware of the parasitic forms because they cause disease. As other organisms have appeared and evolved, parasitic bacteria have adapted along with them, taking advantage of new hosts.

All the bacteria share the characteristic of having no nuclei or other cell organelles bounded by membranes. Reproducing rapidly by **fission,** they generally live in large colonies, though most individual bacteria are **unicellular** organisms.

Two well-known bacterial species are *Escherichia coli* and *Staphylococcus aureus. E. coli* makes up a high proportion of the colon's contents in humans. It rarely causes disease itself, but when found in the environment, it shows that dangerous intestinal pathogens—such as those causing cholera—may be present also.

S. aureus causes boils and other infections; it has been responsible for many hospital-acquired infections among babies and among adult surgical patients. *S. aureus* has adapted well to exposure to penicillin, becoming more resistant to that antibiotic.

Many bacteria have flagella, whiplike tails that can be used for swimming.

Autotrophs have internal membrane systems, called chromatophores, that contain pigments for photosynthesis. Genetic information is located in a single chromosome and (as extrachromosomal DNA) in **plasmids.** Most have a tough cell wall surrounding the cell membrane. In the cytoplasm may be granules or vesicles, and the cell exterior may bear a **capsule,** flagellum, or pili. (Pili are hairlike structures on the surface of bacterial cells.)

Protoctista

During what biologists call the Cambrian explosion, early organisms diversified and radiated enormously. The overall world climate was mild, and there were three periods of land submergence, creating new niches. All protoctists were and are aquatic, but the environment can vary from marine to freshwater to moist soil or tissues. They were limited to specific niches until they could become first more elaborate unicellular organisms and then multicellular organisms.

Organisms evolved in the sea, filling new niches as they appeared. As new organisms evolved, other organisms became capable of preying on them. Many of these early living things died out without leaving any descendants to which modern organisms can be traced. Others led to some modern forms; the numerous Cambrian **trilobites,** for example, seem to be part of the same evolutionary line as modern horseshoe crabs.

Thanks to the brilliant and startling contributions of Lynn Margulis (at the University of Massachusetts at Amherst), the protoctists—the first eukaryotes—are now thought to have arisen as symbiotic bacterial pairings. Different bacteria had different capabilities that allowed them to occupy different environments.

Those changes resulted from a cooperative strategy—organisms with different

Figure 5.4
Some Cambrian Organisms.

abilities combined, forming new organisms with greater repertoires of possible actions. Bacteria were taken up by larger cells and modified to become the membrane-bound structures known as mitochondria or chloroplasts. The cells' nuclei, too, became surrounded with membranes. All of these nuclear or cytoplasmic structures had their own DNA, and the cytoplasmic and nuclear DNA are inherited independently.

Carl Woese has taken the idea of cooperation a step farther. He believes that life began not as one cell complete with genes but as a community of organisms that exchanged genes. The most successful gene combinations endowed their host cells with attributes that led to survival, and eventually three cell lines emerged that became the ancestors of all life—the archaeans, the prokaryotes, and the eukaryotes.[5]

Most protoctists have undulipodia, tail-like formations that they use for swimming; these structures arose in ancestral protoctists even earlier than mitochondria did. Because they superficially resemble the flagella of bacteria, undulipodia are often called flagella, but they are quite different from them in their detailed structure. True undulipodia include the cilia of some protoctists, and even the tails of sperm cells.[6]

There are a great number of possible phyla in the Protoctista. In Table 5.3 the kingdom is shown simply as divided into

Figure 5.5
Symbiosis in Evolution.

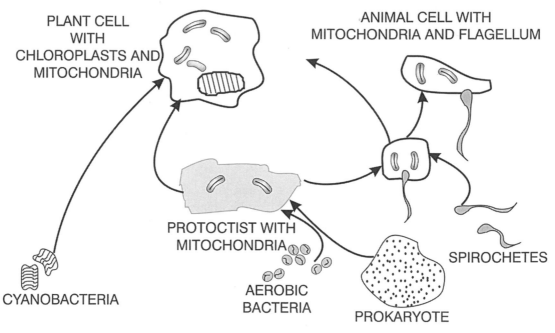

Table 5.3
Groups of Protoctista.

Group	Examples	Characteristics
Heterotrophs	*Trypanosoma, Paramecium*	Parasitic or free-living consumers
Autotrophs	*Euglena*, diatoms, algae	Capable of photosynthesis

two broad categories, based on method of obtaining food and energy.

Heterotrophs

Nearly all animals, fungi, and plants are parasitized by some kind of protoctist. A species of *Trypanosoma,* for example, is a parasite that lives in human blood and causes **sleeping sickness.** It is a flagellate, like *Euglena,* but cannot photosynthesize food; it lives by absorbing nutrients from the blood.

The sarcodines also live by absorbing nutrients from their environment. A well-known sarcodine is *Amoeba,* which has par-asitic and free-living forms. Sarcodines move by extending **pseudopods** and flowing into them.

Sporozoans, or spore-forming organisms, are parasitic protoctists with complicated life cycles. *Plasmodium,* the parasite that lives in human blood and causes **malaria,** is a sporozoan.

Paramecium is one of the ciliate group. Ciliates move by "rowing" many small cilia attached to their bodies rather than by using flagella.

Slime molds, such as *Physarum,* may appear very quickly in places like compost bins, where they absorb nutrients from rotting vegetation. They may form colonies of amoebalike unicellular organisms that give rise to multicellular structures; this duality makes them useful to researchers trying to discover the relationship between unicellular and multicellular life.

Autotrophs

Some protoctists contain chloroplasts and can carry out photosynthesis, as plants do. One of these is *Euglena,* a common pond organism that is both plantlike and animal-like. *Euglena* has a flagellum, like many protoctist flagellates and some bacteria, and can swim about by lashing the flagellum. Autotrophic protoctists also include diatoms, beautiful microscopic creatures that have two-part shells made of silica and live both in salt and fresh water.

Among the protoctists were the first multicellular living things. An organism with many cells has overcome important obstacles to growing larger that inhibit single cells. Because an expanding cell's volume increases at a higher rate than its surface area, it begins to have problems with moving materials in and out efficiently. In contrast, the cells in a multicellular organism have a great amount of total surface area in proportion to their volume.

In addition, the different cells in a multicellular creature can be of different types, as a division of labor is possible for them. Some cells can handle circulation; others, digestion; still others excretion, and so on. As long as materials can move to all cells, specialization and cooperation are possible.

Not all early organisms became multicellular, and the unicellular way of life has its own advantages. But many early organisms did, giving rise to the large living things that are most obvious in the world around us.[7] Trees are not entirely green but are nonetheless autotrophs. They are multicellular and may produce very large leaflike structures.

Green algae may be unicellular, multicellular, or colonial. *Ulva,* or sea lettuce, is one example. Green algae were the ancestors of modern plants. Most of the green algae today are aquatic, but the ancestors of plants may have been semiterrestrial.

Plants

Among the first organisms to colonize dry land were plants, which are generally thought to have arrived on land about 450 million years ago. (Recent evidence from DNA analysis suggests they may have done so as long as 700 million years ago and that they were accompanied by symbiotic fungi.[8]) The bacteria and protoctists still are generally restricted to moist or wet surroundings, but plants evolved in ways that enabled them to move onto dry land during the Paleozoic era or even earlier. For any organism, conserving water was a major challenge to be met in the movement from water to land. Early plants evolved multicellular bodies and anatomical structures that allowed for conserving water and transporting it throughout their bodies.

Today the plant kingdom contains two large groups, the vascular and nonvascular plants. The nonvascular plants are those without conducting tissues (**xylem** and **phloem**), such as the mosses and liverworts. All other plants are vascular. The vast majority of living plants are the vascular group called anthophytes (flowering plants).

Every plant has a life cycle made up of two generations, the sporophyte and the gametophyte. Most of the plants we see around us are advanced species, and we see their gametophytes. The sporophyte generation has been shortened greatly, to the point of invisibility, but it is still a necessary part of any advanced plant's life cycle.

In the bryophytes, both the haploid gametophyte and the diploid sporophyte

TOOLS: STATISTICS AND BIOLOGY

Linnaeus and other early **taxonomists** kept track of organisms' measurable characteristics for comparing and classifying them, but today biologists need more elaborate mathematical methods. One field of mathematics that is often used in biology is statistics.

Determining the statistical probability of a phenomenon or of an event is often important. For instance, the probability of having a tossed coin land heads up is 50%, or a probability of 0.50.

When applied to a population of organisms, probability quickly becomes more complex. Consider the characteristic of middle finger length. In a class of first-year students at the University of Illinois in Chicago, the range of measurements for this characteristic was from 6.1 cm (2.4 inches) to 8.5 cm (3.3 inches). When these individual measurements are plotted as a graph, they form a slightly lopsided bell-shaped curve.

If a curve based on actual data is perfectly bell-shaped, the mean (average) of all the measurements is exactly in the middle, at the top of the curve. This is called a normal distribution. If the area beneath the curve is divided as shown, six marks are produced at equal intervals along the x axis and can be labeled from $-3s$ to $+3s$ (s stands for standard deviation). In a normal distribution the probability that any one measurement will fall within one s on either side of the mean is always 68.3%; the probability that any one measurement will fall within two s on either side of the mean is always 95%; and the probability that any one measurement will fall within three s on either side of the mean is always 99.7%. In the real world, curves showing the distribution of characteristics are seldom exactly bell-shaped.

Some biologists are clarifying the relationships among various plant and animal groups

Figure 5.6
Human Characteristics Graph.

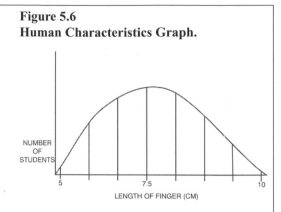

by using statistical methods. They construct numerous complex phylogenetic trees, then calculate and compare their likelihoods of accuracy with a statistical technique called Bayesian analysis. The probability of each branch of each proposed tree is calculated by using data about the organisms' genes. The outcomes with Bayesian analysis can be compared with the trees that were constructed by earlier methods.

In a study by Kenneth Karol (at the University of Maryland) and others, the **charophycean** green algae were shown to be more closely related to land plants than other green algae are. The study also illuminates relationships within the charophycean algae (Karol et al. 2001).

In another study, by William Murphy (at the National Cancer Institute) and others, the phylogeny of living **placental** mammals was studied. The most probable tree shows a major split between North American and South American groups around the time that Africa and South America separated, 101 to 108 million years ago. According to their results, the placental mammals' most recent common ancestor was in Gondwana (Murphy et al. 2001).

Sources: Karol, K., et al. 2001. "The Closest Living Relatives of Land Plants," *Science* 294: 2351–53. Murphy, W. J., et al. 2001. "Resolution of the Early Placental Mammal Radiation Using Bayesian Phylogenetics," *Science* 294: 2348–51.

Figure 5.7
Fern Life Cycle.

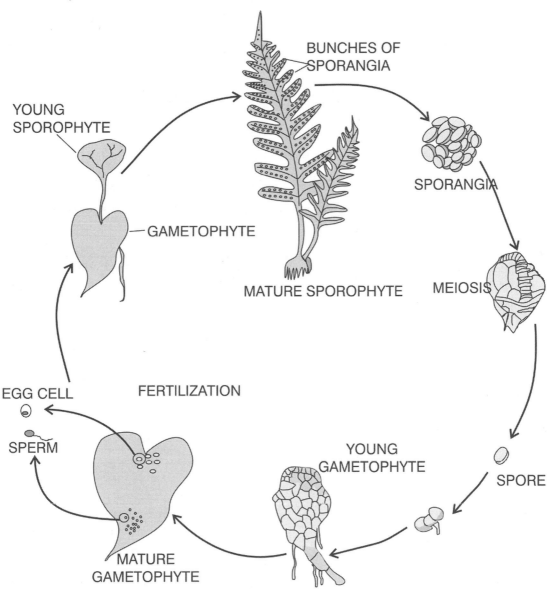

parts of the moss life cycle are obvious. Many primitive plants, such as ferns, have sporophyte generations that predominate over the gametophytes.

The gametophyte, or gamete-producing stage, reproduces sexually. Both male and female gametes may be formed by one plant, or the sexes may be separate, in which case fertilization must be aided by bees or other pollinators. The flowers of angiosperms are the plant's sex organs, and fertilization takes place at the base of the ovary.

The major phyla of the plant kingdom are shown in Table 5.4.

Table 5.4
Phyla of the Plant Kingdom.

Phylum	Examples	Characteristics
Bryophyta (mosses and liverworts)	*Sphagnum* (peat moss), *Marchantia* (liverwort)	Small, nonvascular
Lycophyta (club mosses)	*Lycopodium* (common club moss), *Selaginella* (spike moss)	Vascular
Psilophyta (whisk ferns)	*Psilotum* (whisk fern)	Vascular, but without roots or leaves; whiskbroom appearance
Sphenophyta (horsetails)	*Equisetum* (horsetail)	Tall plants with jointed stems
Pterophyta (ferns)	*Osmunda cinnamomea* (cinnamon fern)	Branching fronds; graceful plants often used for decoration
Cycadophyta (cycads)	*Zamia*	Resemble palms, but have no flowers
Gnetophyta	*Welwitschia mirabilis*	Seeds in cones; plants not tall
Coniferophyta (conifers)	*Pinus* (pine), *Sequoia* (redwood)	Trees and shrubs; seeds in cones; adapted for conserving water
Ginkgophyta (ginkgo)	*Ginkgo biloba*	Branched veins in leaves
Anthophyta (flowering plants)	*Liriodendron* (tulip tree), *Zea* (corn)	Monocots and dicots; flowers, seeds, and fruits in many

Bryophyta (Mosses and Liverworts)

During the Middle Ages, some illnesses were treated with plants that resembled the affected organ. Thus liver disorders were treated with plants we know as liverworts. Their lobed, leaflike structures look something like the lobes of a liver.

All of the nonvascular plants are in this phylum. Though they have structures that look like roots, stems, and leaves, they have no true vascular tissue. Liverworts, mosses, and hornworts appear to have evolved independently from their protoctist ancestors, the green algae.

Because they cannot conduct water and food far, they must remain small and in moist habitats. In those surroundings, water and other materials can diffuse throughout the tissues without having to rise against the pull of gravity. Though they are restricted in their habitats, they continue to occupy them without much competition from plants that have adaptations for living in drier areas.

Lycophyta (Club Mosses)

The club mosses have true vascular tissue—xylem and phloem tubes running through true roots, leaves, and stems. This has enabled them to become tall; in fact, during the Carboniferous period, some lycopods grew as high as 40 m (131 feet). When flowering trees evolved, these lycopods were replaced, but the smaller members of the phylum remain in restricted habitats, such as Olympic National Park in the state of Washington.

Psilophyta (Whisk Ferns)

Though the whisk ferns lack leaves and roots, they are true vascular plants. Their stems contain vascular tissue.

Their evolutionary origin is uncertain. They may be descendants of *Rhynia,* one of the first land plants, or of true ferns.

Sphenophyta (Horsetails)

Horsetails can easily be distinguished from other seedless vascular plants by their stems, which are jointed and have nodes at the joints. In the horsetails, the gametophytic generation has been reduced. One kind of shoot of the sporophyte rises above ground and bears a cone called a strobilus at the top. Within the strobilus, diploid cells undergo meiosis and produce haploid spore cells. A spore is a haploid germ cell that can develop into an adult organism without joining another cell. If a spore is dispersed and germinates in a moist place, it grows into a new gametophyte.

Pterophyta (Ferns)

The ferns are true vascular plants, with roots, stems, and leaves. The undersides of leaves you see may have brown spots, which are clusters of sporangia. Within these are the spores, which can germinate in a moist place and form a new gametophyte. In some species each of the gametophytes is either male or female; in others, both sexes are in the same gametophyte (these species are hermaphroditic). The fertilized eggs

become small sporophytes, which you may know as fiddleheads. These uncoil and grow into large plants, completing the life cycle.

Cycadophyta (Cycads)

The cycads are vascular plants that somewhat resemble palms but they are not flowering plants, as palms are. They bear seeds in female cones, as conifers do, but they are not true conifers either. The gametophyte is microscopic and dependent on the sporophyte. These plants are used extensively for landscapes and interior decoration.

Though most of them are restricted to warm, moist habitats, some have long taproots that enable them to live in deserts.

Gnetophyta

The gnetophytes bear seeds in cones. The *Welwitschia* plant is a huge, sprawling sporophyte seen in deserts. A short portion of the underground stem protrudes above ground, and the leaves and cones protrude from the stem. Apparently its main method of getting water is to collect morning fog in the large leaves.

Coniferophyta (Conifers)

In the conifers, the gametophyte (which may be male, female, or both) is much smaller than the sporophyte. It is found within the cone of the sporophyte. After pollination and fertilization, the diploid zygote becomes a new sporophyte.

Though we associate conifers with the taiga, they are found in other biomes as well. Most are trees, some are shrubs. The majority of them have simple, needlelike leaves coated with a waxy cuticle that conserves water within the leaf. Resin flows through ducts in conifers. The conifers include deciduous and evergreen species.

Ginkgophyta (Ginkgo)

Ginkgo trees are a common sight in city parks. Their fan-shaped leaves have a singular type of venation, with all of the veins branching from the same point. As in cycads, the ginkgo gametophyte is microscopic and dependent on the sporophyte.

Anthophyta (Flowering Plants)

Ranging from grasses to flowering shrubs to trees, the angiosperms are a vast array of species. The two main divisions are the monocots (monocotyledonous plants) and the dicots (dicotyledonous plants). Monocots have seeds that contain just one seed leaf, or cotyledon; parallel veins in their leaves; and other characteristics in common. Dicots have two seed leaves, branching leaf veins, and other shared characteristics.

The angiosperms first appear in the fossil record near the end of the Mesozoic, about 140 million years ago. By 75 million years ago, the huge forests of conifers, cycads, ginkgoes, and seed ferns found in the Mesozoic had largely changed to areas of angiosperms.

The roots of flowering plants anchor the plants in the soil and absorb water and nutrients from it. These materials flow into the root through fine root hairs and pass into the xylem vessels. They are pulled upward through the xylem into the stem and leaves by the force of transpiration (water loss through the stems and leaves). In all parts of the plant, there are also phloem vessels, through which the food made during photosynthesis passes from chloroplast-containing cells to the rest of the plant. The annual rings seen in a cross-section of a tree are xylem; phloem cells are thinner and do not survive the pressure created by a growing tree.

Leaves and stems have openings called stomata, surrounded by guard cells that can open or close the stomata as needed. Air enters through the stomata, and air and water are lost through them.

Most flowers contain both male and female reproductive structures, which are probably modified leaves. The ovary, the part where an egg is formed, is located near the center of the flower, in the base of a vaselike carpel. Above the ovary, a style supports the stigma, or opening of the carpel. One flower may contain one or many carpels.

The entire gametophytic generation lies within the carpel and stamen. There meiotic divisions form the egg and sperm cells. The egg matures within an ovule at the center of the ovary, and sperm cells develop inside pollen grains.

Around the carpel are a ring of stamens, the male reproductive structures. The lower part of a stamen is a filament that ends in an anther. Pollen grains are shed from the anther. When pollen reaches a stigma, a pollen tube grows down to the ovule, where fertilization occurs. The sperm cells are formed within the pollen tube and may fertilize eggs in the same flower or in another one.

**Figure 5.8
Typical Flower.**

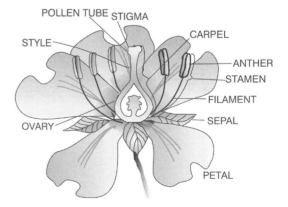

Because **cross-pollination** is necessary for many species, **pollinators** such as insects and hummingbirds coevolved with the flowering plants: The plants became more attractive to insects and birds by developing colors and shapes to which the animals were attracted. Their carpels became sticky, trapping any pollen grains landing on them. The animal pollinators in their turn developed structures for carrying pollen.

Fertilization of the egg leads to formation of a diploid embryo enclosed in a seed. A seed can withstand drying, heat, cold, and other conditions until it is in an environment favoring **germination.** At that time, the tough seed coat can be broken down and the embryo can grow, temporarily using the food stored in the cotyledons. By the time the stored food is consumed, the seedling has begun producing the true leaves and roots that make independent life possible.

The plant tissue that grows and adds to the plant's size is called meristem. It forms new cells throughout the plant's life by mitosis. Meristems are found at the outer ends of stem, roots, and buds. This type of meristem produces growth in length, or primary growth. Secondary growth is growth in diameter, which is brought about by meristem between xylem and phloem in the stem.

Fungi

Fossils of fungi have been found from the Ordovician period, 450 to 500 million years ago. Their ancestry is not yet understood, but they may have descended from protoctists that **conjugated.** Many protoctist groups have evolved to become heterotrophs, have many cells with many nuclei per cell, and reproduce by means of spores; these characteristics define organisms as fungi.

Nearly all fungi are aerobes, and all of them are heterotrophs that characteristically absorb their food, which consists of living or dead tissue. Appearing on the scene before plants, they could take up food from the environment without being able to photosynthesize. Fungi excrete powerful enzymes that break food down into smaller molecules outside the body; dissolved nutrients are then transported into the fungus through the fungal membrane.

Like plants, fungi have two life-cycle stages. In the vegetative phase, a fungus assimilates food for nutrition and growth from its surroundings. In the reproductive phase, it produces and disperses spores that can develop into new fungi.

As parasites, some fungi can harm humans and other animals and plants. The "toxic mold" that sometimes invades homes and sickens the inhabitants is a fungus. So is the organism that infected potato plants and led to the Irish potato famine in the nineteenth century. Dutch elm disease is caused by a fungus, *Ceratocystis ulmi.* Some poisonous species of fungi resemble edible mushrooms and have caused serious illness and death when eaten by mistake.

Other fungi, in contrast, are actually essential for life and health. Penicillin and other antibiotics are the products of the fungus *Penicillium chrysogenum.* Many fungi are important decomposers that break down dead organisms to compounds that can be recycled into new living things.

Fungi are also useful for food production. We eat whole fungi, the mushrooms *Agaricus bisporus. Aspergillus oryzae* can be grown on soybean meal to produce soy sauce. Strains of *Penicillium* ripen cheeses. Other fungi participate in making wine and beer.

The major groups of fungi are shown in Table 5.5.

Zygomycota

The fungus with the unprepossessing common name of dung fungus is actually a

Table 5.5
Major Groups of Fungi.

Group	Examples	Characteristics
Zygomycota	*Rhizopus* (common bread mold) and others	Resemble algae in size and shape
Ascomycota	*Penicillium*; yeasts, morels, truffles, and others	Most form sexual spores in little sacs
Basidiomycota	Mushrooms, puffballs, and others	Most form millions of minute sexual spores; small at base, large at top

beautiful organism that grows only in horse or cow manure. The reproductive phase of *Pilobolus,* as it is more correctly called, has clear bulbs that look as if they have been produced by a glass-blower. At the end of each bulb is a black, caplike mass of spores. To reproduce, the fungus turns so the spores are aimed at the brightest available light, then shoots them toward the light. The spore cases adhere to plants, and if the plants are eaten by an herbivore, the cases are dissolved in the animal's digestive tract, releasing the spores. These can later germinate in the animal's feces and form new fungi.

Pilobolus is typical of the zygomycotes in having spore cases and scattering the spores for reproduction. This fungal group includes *Rhizopus* (common bread mold) and others. Many of them parasitize other organisms.

Ascomycota

Penicillium, yeasts, morels, truffles, and other fungi make up the Ascomycota. Most of them form sexual spores in little sacs.

The ascomycotes and basidiomycotes are more closely related to each other than to zygomycotes and probably are descended from a common ancestor.

Basidiomycota

Like *Pilobolus,* puffballs fire off spores as projectiles. They have evolved a similar strategy to solve the problem of dispersal, but the two groups they represent are not closely related. In the basidiomycotes, the mycelia (threadlike filaments making up a fungus) are organized in distinct layers with separate functions.

Mushrooms, puffballs, and others are members of the basidiomycotes. The mushrooms we see above ground are reproductive bodies that bear spores.

Animals

In spite of their many useful adaptations for survival and dispersal, individual nonanimal organisms are limited in an important way: They cannot move. All their requirements for life must be supplied by their immediate environments. That barrier was crossed by early animals.

The oldest animal fossils are in the **Burgess Shale** in the Canadian Rockies and are about 670 million years old. In adapting to myriad environments over the millions of years since then, animals have evolved many ways of solving problems of survival and reproduction. Animals as a rule can move, they all are many-celled, and they are heterotrophic. These characteristics set

Table 5.6
Phyla of the Animal Kingdom.

Phylum	Examples	Characteristics
Porifera	*Spongia* (bath sponge)	Pores all over body; only one cell layer; only one opening to the body cavity
Cnidaria	*Hydra* (hydra), *Aurelia* (moon jelly)	Jelly-like bodies; two cell layers; many have tentacles; many are radially symmetrical
Platyhelminthes	*Fasciola* (liver fluke), *Planaria* (planarian)	Flat, bilaterally symmetrical bodies; three cell layers; many are parasitic
Nematoda	*Ascaris* (roundworm), *Trichinella* (trichina worm)	Round-bodied worms; many are transparent
Annelida	*Lumbricus* (earthworm)	Worms with segmented bodies
Mollusca	*Helix* (snail), *Mya* (clam)	Soft-bodied, usually surrounded by a protective shell
Arthropoda	*Cambarus* (crayfish), *Cancer* (crab), *Drosophila* (fruit fly)	Exoskeletons; jointed appendages
Echinodermata	*Asterias* (starfish), *Strongylocentrotus* (sea urchin)	Adults imperfectly radially symmetrical (five-part); larvae bilaterally symmetrical; internal skeleton made of calcium carbonate; unique water vascular system
Chordata	*Ambystoma* (salamander), *Crotalus* (snake), *Equus* (horse), *Pan* (chimpanzee), *Squalus* (shark), *Gadus* (Atlantic cod)	Notochord; backbone; most have vertebrae; hollow, dorsal nerve cord; pharyngeal gill slits in embryo and sometimes in adult

them apart from members of the other kingdoms, but within those general attributes, animals are enormously varied.

The best-known phyla of the animal kingdom, arranged in their approximate order of appearance in the fossil record, are shown in Table 5.6. Biologists recognize many other animal phyla and may combine some of these or split them further. Animals are usually divided into two major groups, the vertebrates and the invertebrates—that is, animals with and without backbones. Ninety-eight percent of all living animals are invertebrates.

Invertebrates

Being vertebrates ourselves, humans may tend to think of invertebrates as unimportant. However, they are far more numerous than vertebrates are and have adapted to fill many more niches. Sponges, worms, insects, starfish, and other varied creatures make up animals without backbones.

PORIFERA

Though the poriferans are some of the simplest animals, they are not thought to be

Figure 5.9
Sponge.

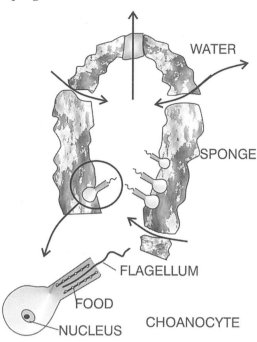

ancestral to any others. Successful in their own aquatic environment, apparently they evolved no further. The sponges arose in the early Cambrian period of the Paleozoic era (about 550 million years ago). Those living today are little changed from many of their Paleozoic ancestors.

The sponges are all sessile (attached to a surface) during their adult life. There they filter food-bearing water in through the thousands of pores that pierce their walls. Food is retained by special cells (called choanocytes) in the sponge's walls, and the water flows out through an opening at the top.

The choanocytes are especially interesting to biologists because they hint at the evolutionary origin of sponges. Because choanocytes resemble choanoflagellates, organisms that swim by lashing their flagella, sponges probably arose from a flagel-

late ancestor. In sponges the flagella swish water and food into the body cavity.

Sponges may reproduce asexually by breaking off fragments that grow into new animals. They also may reproduce sexually: Sperm and eggs may be shed into the water, or the sperm may be transferred to the body cavity of the female sponge, where fertilization and early embryonic development take place.

CNIDARIA

As organisms became larger and multicellular, their cells became specialized to perform various functions for the entire organism—some for digestion, some for passing nerve impulses, and so on. Many cnidarians (jellyfish and related animals) have nerve nets, the simplest nervous systems that connect all parts of an animal. However, the system is decentralized.

Though the two body layers are separated by a substance called mesoglea, which contains contractile fibers and nerve cells, mesoglea is not a true cell layer. True cell layers give rise to various specialized tissues. In all animals that have three cell layers, the embryo's ectoderm gives rise to the skin and nervous tissue; endoderm forms the linings of all tracts, such as the digestive and urinary tracts; and mesoderm produces muscle and bone as well as various internal organs. Cnidarians have no real organs.

The cnidarians are radially symmetrical, gelatinous animals. The body may be tall and cylindrical—a form called a polyp—or a bowl-shaped form called a medusa. These are different stages in the organism's life cycle. The medusa can swim, but the polyp is attached at its base to a substrate. A colony of one species may contain both polyps and medusae.

Medusae release sperm and egg cells into the water, where fertilization takes place. The resulting embryo grows into a polyp. Though this life cycle somewhat resembles the alternation of generations in plants, both medusae and polyps are diploid. (Only the gametes are haploid.)

Like poriferans, cnidarians have a central digestive cavity with only a single opening to the outside. However, they evolved structures that increased their predatory ability over that of sponges. Tentacles surrounding the mouth have special cells that sting prey. After prey has been passed through the mouth into the central cavity, digestive enzymes are secreted into the cavity, and extracellular digestion begins. It is limited mainly to partial breakdown of proteins. When the food has partly disintegrated, the cells lining the cavity take up the fragments, and digestion continues within food vacuoles inside the cells.

PLATYHELMINTHES

The first animals to have bilateral symmetry were the flatworms. This characteristic helped make it possible for early organisms to develop head ends, necessary for the eventual evolution of brains. An animal with a head can better meet the environment it is entering, where predators or prey may be waiting. Mesoderm, a true cell layer, also first appeared in this phylum. It separates the ectoderm and endoderm layers.

Flatworms probably appeared in the oceans around 570 million years ago, but few fossils have been found to substantiate this theory. Not having bones or other hard parts, they left little evidence of their existence.

Many flatworms have central cavities like those of cnidarians, though otherwise their bodies are more complex. In planarians, for instance, the mouth opens into a tubular chamber called the pharynx. The pharynx leads into a cavity that branches throughout the planarian's body. As in cnidarians, some extracellular digestion occurs in the planarian gastrovascular cavity, with the small food particles then being engulfed by gastrodermal cells and digested intracellularly. The additional process of extracellular digestion frees cnidarians and flatworms from exclusive reliance on intracellular digestion.

Flatworms have a great ability to regenerate lost parts, as they retain undifferentiated stem cells in their bodies throughout life. If a flatworm is cut into hundreds of pieces, each one will regenerate a new flatworm from the stem cells.

NEMATODA

Most animals that are more advanced than cnidarians and flatworms have a complete digestive tract, consisting of a tube with two openings—a mouth at one end and an anus at the other. This arrangement makes possible a more efficient system, with a series of steps carried out in separate, specialized sections. One section may be specialized for breaking down food mechanically, another for temporarily storing it, a third for digesting food with enzymes, and a fourth for absorbing the products of digestion. Evolution has modified these divisions with the result that different animals can inhabit a variety of niches.

Nematodes are nearly everywhere. Parasitic species are so common that nematode biologist N. A. Cobb (1859–1932) once said, "If all the matter in the universe except the nematodes were swept away, our world would still be dimly recognizable." They are found as parasites on plant roots, in animal intestines, and in much of the rest of the biosphere. Free-living nematodes also are found nearly everywhere: 20,000 species have been described.[9]

ANNELIDA

Annelids are found in all sorts of environments, including soil, fresh water, and sea water. They feed by preying on other organisms or scavenging dead materials.

The fossil record of annelids is limited because they have almost no hard body parts. Based on the evidence of tubes, jaws, and teeth, biologists think the annelids probably evolved in the sea beginning about 600 million years ago.

The larval form (called the **trochophore** larva) of **polychaete** annelids may have descended from an ancestral flatworm. Another group of annelids, the **oligochaetes** (which include the familiar earthworms), probably developed from polychaetes and leeches, the third major group, from the oligochaetes.

Polychaetes are called paddle-footed worms because of their many flat "feet" that help in movement and breathing. They turn over thousands of tons of sand on the sea floor as they burrow through it.

All annelids are characterized by external rings that correspond to internal separations between compartments. Within each of the compartments some parts—nervous, muscle, and excretory organs, for example—are repeated. Others are located in only certain segments. In annelids there is the beginning of a brain, the cerebral ganglia, at the head end. Some have eyes and other sense organs. During embryonic development of these worms, the mesoderm develops a cavity called the coelom, which is characteristic of all advanced animals.

Earthworms aerate soil as they move through it and enrich it with their castings. They are so successful at this that in India earthworms have been used to restore degraded soil on estates where tea has been grown for long periods.[10]

Even today, leeches are sometimes used for bloodletting; their saliva contains an anticoagulant and an anesthetic, providing an effective combination for controlling swelling following some kinds of surgery.

An advance in excretion can be seen in the annelids, which have ciliated structures called nephridia. Found in most of the body segments, nephridia draw in dissolved wastes from the coelom's fluid and absorb other wastes from blood vessels, then discharge them through external pores on the body surface.

MOLLUSCA

The fossil record gives little clue as to how the mollusks originated; they appeared in Precambrian times. They are not primitive, however: Like annelids, mollusks have a coelom surrounding a complete digestive tube. They have a circulatory system and special structures for respiration as well.

The group is characterized by a body covering, the mantle, that in many organisms secretes a hard shell. In aquatic mollusks the mantle pumps water in and out of the shell, carrying food, oxygen, and wastes with it. The shell can be opened by muscles, allowing the animal to provide protection for itself or to feed—a useful adaptation for these soft-bodied creatures in an environment where either predators or prey may be present.

ARTHROPODA

Among the early arthropods were arachnids and centipedes, the first clearly terrestrial animals. Fossils have been found from 420 million years ago, and it is likely that earlier animals had already moved onto land.

Arthropods solved not only the problem of moving onto land but also the more daunting challenge of flying into the air. Some insects evolved wings, allowing them to radiate more easily into new environments and move around in their home area. As a result, today arthropods are by far the most diverse and numerous of the animal phyla.

In occupying various niches adjacent to humans, arthropods have proved both a bane and a blessing to humankind. They carry the bacteria or protoctists that cause many serious diseases; they compete with us for grain and other foods; they may bite us or live quietly on our scalps. On the other hand, they carry pollen for our fruit trees, and some arthropods feed on other, harmful varieties.

The first fossil arthropods, including some trilobites and crustaceans, are seen in rock layers from the Cambrian period (570 to 500 million years ago).The arthropods today include crustaceans, chelicerates, and mandibulates, groups that are given the status of separate phyla by many biologists because they appear to have separate evolutionary lineages.

In many ways the arthropods are like the annelids: Both are segmented, and the polychaete annelids have a pair of appendages on each segment, as the arthropods have. The nervous system plan is similar in the two groups. Also in both groups is a tubular, dorsal heart, which is lost or modified in some. Annelids and arthropod embryos have a coelom, though arthropod adults do not.

Mandibulates, which include the insects, contain more species than any other animal group. (In all, more than 30 million species of animals are estimated to exist now.) The bodies of mandibulates have three parts—the head, thorax, and abdomen. The bodies are covered with a cuticle, made of chitin and protein, that provides the animals with a tough exoskeleton for support and protection.

Mandibulates' respiration is aided by air passing into the body through spiracles, pores in the cuticle. It flows into tracheae (thin-walled tubes that branch throughout the body and provide gas exchange to all parts of it). Thus these organisms were able to breathe air before larger animals evolved lungs for that purpose.

ECHINODERMATA

This group has members with spiny skins, including the starfishes, sea cucumbers, and sea urchins. It has also members with smooth coverings, such as the sand dollars. Every echinoderm, however, has three unique features: five-part symmetry, an internal skeleton made of calcium carbonate; and a divided coelom that includes a water vascular system. This system consists of canals that radiate into the arms or other body sections, carrying sea water. The canals protrude through the body wall as tube feet, structures that are adapted for collecting food, acting as **chemoreceptors,** secreting mucus, and respiring, as well as for moving.

Because the phylum Echinodermata was already well diversified by the early Cambrian period, much of its evolution must have taken place in the Precambrian. Echinoderms are believed to have descended from bilaterally symmetrical ancestors. Early echinoderms were adapted to life on the sea floor surface, just like sponges and many mollusks are. Today many species are adapted to the rigorous life of the intertidal zone and are often seen in tidepools. All are marine organisms.

Chordata

Embryos of chordates, the phylum that includes humans, all have a long stiff rod called a notochord extending along the body. The notochord supports the body of the embryo. In a few of the more primitive chordates, such as lancelets and sea squirts, the notochord is retained throughout life. (Thus these animals are invertebrates.) In

**Figure 5.10
Lancelet.**

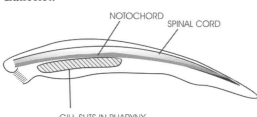

most groups, however, it is eventually replaced by cartilage or bone.

The embryonic notochord is one of three distinctive characteristics of chordates. The other two are a hollow, dorsal nerve cord and pharyngeal gill slits in the embryo (sometimes in the adult). Because they demonstrate the three chordate characteristics very clearly and simply, lancelets (amphioxus, or *Branchiostoma*) are often studied by beginning zoology students.

Chordates arose sometime before 590 million years ago—that is, before the fossil record began. Like annelids, the first chordates were soft-bodied, so they left few fossils for study. Based on embryos, the phylum Chordata is now believed to have evolved from the echinoderms. The earliest chordates may have looked like modern sea squirts.

The first chordate groups to evolve had no vertebrae. Today some primitive invertebrate chordate groups remain, but most chordates have vertebrae as adults. All of the vertebrates are often grouped together in the subphylum Vertebrata. The large record of vertebrates began around 400 million years ago; because this group had bony skeletons, they left many fossils.

All vertebrates have a nerve cord surrounded by a vertebral column and a brain enclosed in a cranium. The major classes, shown in Table 5.7, have other distinctive characteristics.

AGNATHA

The earliest fishes (the agnathans) had no jaws. Like *Branchiostoma* and other early chordates, from which they probably evolved, they have a notochord running along the dorsal side of their bodies.

The earliest vertebrate fossils of agnathans are from the Ordovician period in North America, more than 500 million years ago. The North American jawless fishes may have lived in shallow coastal marine waters, where their remains became fossilized; or they may have been freshwater organisms that washed into coastal deposits. In either case, they left fossils because they had an armorlike heavy dermal covering.

Today only the hagfish and the lampreys represent the agnathans. They have round mouths that can fasten onto prey so the fish can feed without being dislodged and even tunnel into the prey's viscera. No fossils are known of their more primitive ancestors. In fact, remains of the earliest vertebrates may never be found.

CHONDRICHTHYES

The jaws we associate with sharks appeared in evolution as a modification of the **gill arches** and had the enormous advantage of allowing their owners to feed on large prey animals. That allowed the jawed fish to grow larger themselves than the agnathans had been.

Like the agnathans, the cartilaginous fish have cartilaginous skeletons. However, both groups are descended from ancestors with bony skeletons.

Around the beginning of the Devonian period, 410 million years ago, a diverse group of fishes, the placoderms, arose. They had paired fins, jawlike structures, and bony skeletal tissue. Very successful during the Devonian, they died out at its end. The placoderms gave rise to two

Table 5.7
The Classes of Vertebrates.

Class	Examples	Characteristics
Agnatha	*Petromyzon* (lamprey), hagfish	Jawless fishes
Chondrichthyes	Shark, skate	Skeleton made of cartilage
Osteichthyes	Perch, trout	Skeleton made of bone
Amphibia	Frog, salamander	Thin skins, usually without scales; some have lungs; many metamorphose into adult forms
Reptilia	*Alligator* (alligator), *Heterodon* (hognose snake)	Usually have four legs; dry skin covered with scales; amniote eggs; exothermic
Aves	*Archaeopteryx* (fossil bird)	Feathered; bodies adapted for flight; endothermic
Mammalia	*Delphinus* (dolphin)	Nurse young; hair on body; endothermic; most bear live young; most have placentas

groups of descendants, the osteichthyans (bony fishes) and the chondrichthyans (cartilaginous fishes).

It was once believed that the cartilaginous skeleton evolved earlier than the bony skeleton because cartilage is formed as a precursor to bone. Now it seems more likely the cartilaginous fish arose from osteichthyan ancestors. The chondrichthyans arose in fresh water, but moved into the sea. To cope with the saltwater environment and maintain their equilibrium, they had to evolve a new kind of kidney for retaining salts. All the fishes—cartilaginous and bony—have a type of kidney that is more advanced than those of earlier groups. It contains **renal corpuscles** that allow the excretion of water and reabsorb salts.

The cartilaginous fish radiated and thrived during the Mesozoic but became greatly reduced in the Triassic, which was a drier period. Then, in the Jurassic (about 200 million years ago), the seas enlarged again, and the ancestors of modern sharks, skates, and rays appeared.

OSTEICHTHYES

By the late Devonian period, most kinds of fish had lungs. This gave them a selective advantage in times of drought, when other fish could not survive.

Today many descendants of the lunged fishes have swim bladders that are used for buoyancy or for producing sound; these bladders are modifications of the lungs in lunged fish. Other lunged fishes gave rise to modern lungfish, which can live in water that has very little dissolved oxygen.

The Osteichthyes group includes 20,000 species and more than 400 families of modern bony fishes. Their main characteristic is a skeleton composed at least in part of true bone rather than cartilage. Osteichthyans also have a swim bladder, plate-like bony scales, gill covers over the gill chamber, and a bony skull. They fertilize their eggs externally.

Osteichthyans live in all kinds of freshwater and ocean environments, including deep-sea habitats, thermal springs, and even the water in caves. They have a great variety of shapes and behaviors, also, with body sizes ranging from 12 mm (0.5 inch) for the pygmy goby *(Pandaka pygmaea)* to 4.5 m (15 feet) for marlins and swordfishes.

AMPHIBIA

Though lunged fishes could breathe, to move onto land, animals had to evolve limbs for walking. The transitional form *Eusthenopteron* and its relatives had rudimentary limbs and dragged themselves out of the water, becoming ancestral to early amphibians.

Amphibians never freed themselves completely from the water, however. Today most amphibians begin life as aquatic eggs and larvae, then undergo metamorphosis into adults that can live on land. (The change from tadpole to frog is a familiar example of amphibian metamorphosis.)

Today there are three groups of amphibia—the salamanders, the anurans (frogs and toads), and the caecilians—which vary greatly in size and structure. Their total length varies from 27 mm (1.06 inches) to more than 1.5 m (5 feet). The salamanders have long tails, and two pairs of limbs of about equal size. Anurans have long hind limbs and the large eyes by which they are easily recognizable. The long, slender, limbless caecilians are adapted for burrowing and somewhat resemble snakes.

Amphibian eyes show advances over the eyes of fish, having lids, associated glands and ducts, and muscles that allow **accommodation** and depth perception. They have true color vision: Green rods in the retina, allowing perception of many wavelengths, are found only in amphibians. Such modifications were useful for capturing prey in the animals' new terrestrial habitat. The amphibian auditory system is also specially adapted for receiving both sound waves and **seismic** signals.

Selection by the terrestrial environment has modified the anatomy and behavior of various amphibians. One problem in the new environment (which lacked the support provided by water) was stabilizing the head on the **axial skeleton.** Salamanders met the challenge by evolving an extra set of articulations with the vertebral column. Other amphibians developed specialized trunk muscles that accomplished the same thing.

A part of the cranium, the hyoid apparatus in the floor of the mouth, makes it possible for salamanders to capture prey by rapidly flicking and retracting their fleshy tongues. They hold and manipulate food in the mouth, using the teeth and tongue. (Most salamanders have no strong muscled jaws or specialized teeth.)

Salamanders have well-developed eyes and nasal organs. Because salamanders do not depend on vocal abilities, their auditory apparatus is less specialized than that of anurans, whose croaking calls are well known.

Most caecilians live underground and have adaptations in shape for burrowing, which makes them look something like snakes. They cannot extend their tongues as far as anurans and salamanders can. Their vision is poor, but their olfactory and chemosensory abilities are good. Caecilians probably hear less well than salamanders and anurans. They move through soil by what is called concertina locomotion—the body alternately folds and unfolds its entire length in an accordionlike movement.

More numerous, widespread, and diverse than either salamanders or caecilians, anurans have become specialized for inhabiting many different environments.

Their long, powerful hind limbs rapidly push the fused head and trunk forward. Many can jump for impressive distances, as noted by Mark Twain in his short story, "The Celebrated Jumping Frog of Calaveras County."

Many anurans can also burrow or climb trees, having evolved certain changes in limb proportions and joints. Arboreal (tree-dwelling) anurans have long limbs and digits that end in large, adhesive pads. Burrowing anurans, in contrast, have short, sturdy limbs; their feet have a spadelike shape.

Frogs and toads need good vision for feeding and movement, so their eyes have become large and well developed. Their ears are highly specialized also because they vocalize as part of their mating and territorial behavior. Most species of anurans have an external tympanum (eardrum).

The anurans' many adaptations have allowed them to diversify into a great variety of habitats. However, during the late twentieth century their numbers began diminishing, partly as the result of losing appropriate habitats for egg-laying and early development. Pollutants may have deformed or killed some anurans also. Introduced species may have competed with them, and in some cases they have been commercially exploited by humans.

REPTILIA

Living reptiles include snakes, turtles, lizards, crocodiles and alligators, and tuatara. Extinct forms were animals such as the dinosaurs, pterosaurs, and ichthyosaurs.

Reptiles are intermediate in evolutionary advancement between amphibians and warm-blooded vertebrates (birds and mammals). They completed the transition to terrestrial life begun by the amphibia, breathing air and laying **amniote** eggs that provide a watery immediate environment for the developing embryo. Reptiles have internal fertilization and scaly bodies.

When a reptile hatches, it is ready to carry out all the activities of the adult, except for reproduction, and it already has a body adapted for terrestrial life. This is made possible by a long embryonic period and a large amount of yolk in the egg.

Reptiles first appear in the fossil record of the Carboniferous period, more than 280 million years ago. By the Triassic, about 50 million years later, they began to dominate the land. That continued through the Mesozoic era, 65 to 225 million years ago. Reptiles managed to adapt to nearly every kind of terrestrial environment—deserts, swamps, forests, grasslands, rivers, and lakes, as well as the air and the oceans. Most reptilian groups became extinct at the end of the Mesozoic, however, and the mammals expanded to dominate the land.

The reptiles' move from water to land as a lifetime environment required two basic changes. The first change was in the skin. Modern amphibians have naked skins without tough scales or hair, though the caecilians do have small, fishlike scales embedded in the skin. These amphibian scales are too thin to protect against drying out, so amphibians must remain in water or in very humid places, emerging onto land only temporarily. In reptiles there are thick keratin (or horn) scales in the outermost skin layer that can prevent water loss.

Another major change in the reptiles was the development of the moisture-retaining shelled amniote egg, which made it possible for reptiles to exploit most terrestrial environments. Reptile eggs can be laid wherever there is some air—in deserts or forests, on grass or sand. Because a reptile egg must be fertilized before the shell is deposited around it, internal fertilization became necessary in reptiles.

Having scales as armor and an egg that was free from water, reptiles could expand over most of the Earth. That led to evolution both of the many types of reptiles and of the birds and mammals.

One reptile, the pterosaur, also took to the air as a flying dinosaur. Instead of true wings, pterosaurs had huge flaps of skin that stretched from their flanks to their fingers.

AVES

Birds are easy to distinguish, being the only animals with feathers. Most biologists think now that birds are descended from the theropods, dinosaurs that had two legs and could run. Birds are warm-blooded, reptiles are not; some of the important evidence connecting the dinosaurs and birds indicates that dinosaurs, too, were warm-blooded.

The oldest known real bird fossil is that of *Archaeopteryx,* which lived 150 million years ago. It had many features like those of modern birds (including feathers) but also had a long, reptilian tail.

After the demise of the dinosaurs and before large mammals succeeded them, birds dominated Earth for a time. Some were huge, having wingspans of more than 6 m (20 feet). During their evolution, birds have become very specialized for flight. In addition to feathers, they have a unique one-way breathing system, hollow bones that are light yet strong, a simplified skeleton, and strong flight muscles.

Having the ability to fly where no other animals could, birds were able to radiate into many environments and evolved adaptations for a variety of conditions. As a result of selection, some modern birds can live in extreme climatic conditions. They also evolved a huge assortment of beaks that enabled them to eat different seeds, nuts, insects, and other foods.

MAMMALIA

Like the birds, mammals are warm-blooded. They also have easily identifiable characteristics that belong to mammals alone: They have hair and nurse their young. The mammary glands that produce milk probably arose as modifications of the glands associated with hair.

Hair was a useful adaptation for animals that needed to adjust to changes in environmental temperature. The reptiles that preceded mammals had to solve the temperature problem by changing their behavior to avoid becoming too warm or cold. Mammals, however, evolved hair and other structures that helped them maintain a constant warm body temperature.

Mammalian circulatory systems are also more complex than earlier groups', keeping venous and arterial blood separated and providing highly oxygenated blood to the complex brain. Along with evolving these changes, mammals developed limbs that projected underneath the body, not to the side as reptiles' do. This made it possible for them to move quickly as predators or as prey.

Reptiles lay eggs and do not care for their young, but mammals began producing live young and retaining them inside the mother's body until they were well developed. In most mammals maternal care continues well after birth. The more advanced mammals educate their young in addition to caring for them physically.

Though all these advances were helpful to the earliest mammals, the reptiles were dominant because of their sheer physical size until 65 million years ago, when an asteroid's impact (hypothesized) or some other catastrophic climatic event killed off many of the largest reptiles. At that point, the mammals had less competition and could begin their takeover of the land.

APPLICATION TO EVERYDAY LIFE FEATURE: ZOOS, PRO AND CON

Nearly everyone enjoys spending a day at the zoo. Zoos have been popular throughout history, and today about half a million vertebrates can be found in zoos throughout the world.

In zoos, animals from all over the world can be observed interacting with each other—and often with the human visitors as well. You can see how live animals behave, feed, and move much better than you can see them in even the best movie. If you become especially interested in one species, you may be able to buy or borrow books or videos with further information about it.

In modern zoos, the animals are usually kept in simulations of their natural surroundings, where they may have large spaces in which to roam. One of the first to exhibit animals in their natural surroundings was the San Diego Zoo. In that zoo's Ituri Forest, visitors can see a simulated African rain forest containing forest buffaloes, hippos, otters, okapi, and monkeys. In other parts of the zoo, simulated environments with native fauna are called the Polar Bear Plunge, Tiger River, Gorilla Tropics, and Sun Bear Forest. The Polar Bear Plunge is an icy arctic tundra, with Siberian reindeer, arctic foxes, polar bears, northern birds, and native plants. Visitors can watch the animals from underground viewing areas, and a Web cam makes it possible to watch them from anywhere on the Internet (visit www.sandiegozoo.org).

Biodiversity (the diversity of organisms and ecosystems in the natural world) is conserved by zoos that serve as research centers, increase public awareness of threats to species, and breed and reintroduce endangered species. Several species, including the California condor *(Gymnogyps californianus)* and black-footed ferret *(Mustela nigripes)* can be found only either in zoos or as species recently released from zoos. The peregrine falcon *(Falco peregrinus)* and the alpine ibex *(Capra ibex ibex)* have been reintroduced from zoos to portions of their range.

Yet many people object to the idea of keeping animals "in cages" rather than in their natural habitats. In the past, many zoo animals were caged and otherwise poorly cared for, even treated cruelly, providing much fuel for the arguments against zoos. Even if the animals were fed enough nutritious food and given enough exercise, they lacked the stimulation and variety they needed. If too few other animals were available for interaction, their behavior sometimes became neurotic or at least atypical of normal behavior. Opponents of zoos argued that it would be better to conserve the animals in their native habitats.

Zoos are supposed to be educational as well as entertaining, and the best zoos do provide educational materials as background for the exhibits. In some cases, though, zoos can miseducate. In an article he wrote about zoos, one wildlife photographer protested about the birds he found in two lakes at a zoo in northern California. One represented Lake Maracaibo, in Venezuela; the other, Lake Victoria, in Africa. The photographer found that many of the birds were not representative of those lakes at all, so the exhibit was actually miseducational. In the Lake Maracaibo area he found Chiloe widgeons, black-necked swans, and rosy-billed pochards, South American species that are actually restricted to countries south of Venezuela. In the Lake Victoria area he found a red-breasted goose that is native to northern Siberia, and flamingoes belonging to a New World race (which are sometimes considered a different species from the species found in Africa). The sole value of these lakes seemed to be as an entertaining and attractive display. (Note: The zoo has undergone renovation since then, so the situation may have improved.) (McKay 2002)

There are limits in space and costs to how many species can be conserved by zoos; nevertheless, zoo managers have calculated that zoos can sustain small populations of about 900 species. In addition, after a few generations in captivity, some species can be returned to their native habitats. Species being

CONTINUED

managed in this way include the Guam rail *(Rallus owstoni)*, the Puerto Rican parrot *(Amazona vittata)*, and the Mauritius pink pigeon *(Nesoenas mayeri)*.

In answer to criticism, zoo managers have made enormous improvements in zoos over the past 40 or 50 years. Large groups of well-treated animals can now be seen in spacious, natural environments. Over time the conservation of native habitats has become more difficult. If an area is being overrun with poachers, as has happened in Africa, the threatened animals may be safer (if not happier) in the artificial setting of a zoo on another continent. If an animal species is threatened with extinction, zoos may be the only means of maintaining the species' genome.

Finally, the phenomenon of ecotourism has led to problems with the few wild places left on the Earth being overrun with visitors. In some cases it may be better to keep the human visitors at home, seeing a great variety of animals in zoos, than to ruin the animals' natural habitats.

Even so, many people still oppose the idea of zoos altogether. Some argue that the money spent on zoos can be spent instead on programs that protect wildlife populations and their habitats, because endangered species ultimately are more likely to survive in their own habitats than in zoos. They may also argue that animals born in captivity or rescued from unsafe environments can be kept in nonprofit animal sanctuaries.

Source: McKay, Barry Kent. 2002. "When Zoos Tell Lies," Animal Protection Institute Web site, www.api4animals.org (accessed September 1, 2002). Also personal communication (e-mail) from Barry Kent McKay.

There are three major groups of living mammals. The first (which includes humans) is made up of the placental mammals. Marsupials are another order and include all of the pouched animals, such as opossums and kangaroos. The third order, the monotremes, are much rarer than the other groups. They have the diagnostic characteristics of mammals, but they lay eggs and do not give live birth. Monotremes include echidnas and platypuses.

Insectivora The first mammals to hide from the dinosaurs and then emerge as their successors were probably similar to modern insectivores. This group includes small animals, such as shrews, moles, and hedgehogs. As their name implies, they eat mainly insects, but they may also eat worms and mollusks.

The insectivores have poor vision, clawed fingers and toes, and small brains. Their senses of smell and hearing are keen.

Many of them burrow through soil, an adaptation that would have helped them avoid detection by reptilian predators. They are found throughout North and Central America, Europe, Asia, and Africa.

Chiroptera This order is made up entirely of bats. Like the insectivores, many chiropterans eat insects. Other bats have evolved in another direction, specializing in eating fruit. Today bats have radiated into virtually every region in the world except for the polar regions and some remote islands.

Bats are the only mammals that can truly fly. Though basically blind, they are aided in navigating by an echo-sounding system; bats send out beams of supersonic squeaks that come back to them as echoes. Using the echoes to determine where their prey are located, bats can find their food without being able to see it.

Primates Some early insectivores were the ancestors of the primates. The first pri-

Table 5.8
Orders of Placental Mammals.

Order	Examples	Characteristics
Insectivora	Hedgehogs, moles, shrews	Poorly developed eyes, unspecialized teeth, small brain
Chiroptera	Bats	Web of skin connects hind and front limbs; another connects fingers
Primates	Lemurs, monkeys, apes, human beings	Most have nails rather than claws; many teeth; eyes face forward
Edentata	Sloths	No front teeth
Pholidota	Pangolin	No teeth; hair modified to form scales that cover body
Tubulidentata	Aardvark	"Nails" on toes are intermediate between claws and hooves; few teeth in adults
Rodentia	Porcupine, squirrel, rats	No canine teeth, broad molars, chisel-like incisors
Lagomorpha	Rabbits, pikas	Rodent-like teeth; four upper incisors; very short tail
Cetacea	Whales	No hind limbs; front limbs modified as flippers; small eyes in very large head; no hair on adults
Carnivora	Lions, dogs	Small incisors, premolars adapted for shearing; sharp claws
Proboscidea	Elephants	Modified upper incisors are tusks; herbivorous; modified nose and upper lip form a trunk
Sirenia	Manatees, dugongs	Herbivorous; marine; no hind limbs; few hairs; broad, flat tail
Perissodactyla	Zebra, rhinoceros	Odd number of toes, modified as hooves; herbivorous; well-developed molars
Artiodactyla	Hippopotamus, peccary	Even number of toes, modified as hooves; herbivorous; complex stomachs; many have antlers or horns

mates—in the Paleocene epoch, about 60 million years ago—lived in trees and were active and diurnal, in contrast to the insectivores. They evolved large eyes, enabling them to fill niches unavailable to insectivores or chiropterans. Over time, many primates have evolved further, leaving the trees and becoming adapted to life on the ground as the climate became cooler and drier. They have shorter arms than earlier primates, for instance, and humans walk upright. All primates have separate, mobile fore limb bones (radius and ulna) and hind limb bones (tibia and fibula). Their feet have five toes.

The primates known as prosimians include such animals as lemurs, lorises, tarsiers, lorises, and bushbabies. Some of these are nocturnal.

More advanced primates are called anthropoids because they more closely resemble humans than any other animals do. Anthropoids, which comprise monkeys, great apes, and humans, have larger brains and flatter faces than prosimians. In addition, the olfactory part of the brain is smaller and the cerebrum is larger.

Nonhuman primates are found today in Central and South America, Africa, Madagascar, and southern Asia.

Edentata This group is mainly confined to the tropical areas of South and Central America. It includes sloths, armadillos, and anteaters, all animals having very large claws and small brains. They have two or three long fingers on each hand. Though the word *edentate* means "toothless," only the anteaters in this group have no teeth at all. Other edentates are missing some teeth or have very simple teeth.

Two groups of edentates—glyptodonts and giant ground sloths—became extinct. The glyptodonts were heavily armored and somewhat resembled their smaller modern armadillo relatives. The giant ground sloths were the size of small elephants, weighing several tons.

Pholidota This order, the pangolins or spiny anteaters, includes only seven living species. Pangolins have soft, hairy underbellies but otherwise are covered with large scales of armor. These animals have no teeth and small brains. They are sometimes said to look like pine cones with legs. Pangolins feed on burrowing insects, such as ants and termites. They are found in tropical Africa, southern Asia, Sumatra, Borneo, Java, and the Philippines.

Tubulidentata Only one species, the aardvarks, represent this order. They are found only in Africa, south of the Sahara. The aardvark is probably the most primitive living protoungulate (earliest hoofed mammal). Little is known about the ancestry of the group; their closest living relatives are apparently the elephants, hyraxes, and sirenians. They were once thought to be closely related to pangolins, but the similarity probably resulted from convergent evolution of the two groups.

They do not have true hooves but structures that are intermediate between claws and hooves. These enable the pig-sized animals to burrow and to tear open termite mounds, which are very hard. Aardvarks have long, extensible tongues with which they feed on termites, other insects, and plants. Their cheek teeth have a tubular shape that gives the order its name. No canine or incisor teeth are present.

Rodentia Rodents are the largest order of mammals, including about 1,500 living species. Rodents remained small as they evolved but have spread into an enormous variety of environments throughout the world because of their adaptability. They are native on all continents except Antarctica. They can swim, leap, hop, climb, burrow, and run, depending on what the situation requires. In addition, rodents have teeth (a single pair of incisors in each jaw) that are specialized for either gnawing or chewing, and muscles control their jaws so that the teeth not being used are not worn out and the animal doesn't swallow clumps of soil or wood chips. The largest living rodent is the capybara, which is as large as a pig.

Squirrels, chipmunks, and other animals are found in one group. A second group is made up of rats and mice. Porcupines, cavies, agoutis, and their relatives are in the third group.

Lagomorpha Rabbits, hares, and pikas make up the lagomorphs. All members of this order have the same general body shape and short tails, but the pikas have shorter ears and limbs than the rabbits and hares. Their habitats differ, also—the pikas living in rocky areas and the rabbits and hares in grasslands. Although lagomorphs somewhat resemble rodents and have similar chisel-like teeth, the numbers and types of teeth are different. All lagomorphs have long, soft fur.

Lagomorphs are related to the rodents no more closely than to any other order of mammals. They are often preyed on by carnivores, but their high rate of reproduction keeps their population size large.

These animals are found in nearly every part of the world; though not native to the Antarctic or the Australia–New Zealand areas, they were introduced to Australia and New Zealand in the eighteenth century by European explorers, with devastating consequences. Not having natural competitors, the rabbits quickly spread through those countries and became important pests. Ever since, farmers and wildlife managers have tried a variety of ways of controlling the rabbits' numbers—shooting, poisoning, destroying rabbit habitat, fencing, fumigating, even introducing contagious diseases to kill them. No method has been very successful.[11] Not too surprisingly, New Zealand rabbits are exported for sale as pets.

Cetacea The cetaceans arose from land mammals that lost their hind limbs and otherwise became completely adapted to aquatic life. The group includes dolphins, porpoises, and whales, which fall into two broad groups—toothed whales and whalebone whales.

Toothed whales feed on smaller marine mammals, large shellfish, and fish. Whalebone whales have plates of baleen in their mouths that filter out large numbers of very small organisms in marine plankton, making it possible for these enormous whales to subsist on very small prey.

The great blue whale is the largest of all mammals, weighing more than 100 tons. Probably no other animal that large has ever lived on Earth.

Cetaceans have no neck or external ears. Toothed whales have a single nasal opening (blow-hole) on top of the head; whalebone whales have a double opening. Their hair has been reduced to a few hairs or none. All cetaceans have a streamlined shape, paddlelike front limbs, and vestigial hind limbs.

Whales can communicate for miles in the ocean by vocalizing. Humpback whales, especially, are known for their complex songs. They also use a sonar system for navigation by echo-sounding, much as bats do.

Whales live in the water for their entire lives unless they are beached by a storm or illness. They cannot live out of the water for more than about 24 hours because of their tremendous weight; their internal organs are crushed, and they die. Many of them can dive to great depths in the ocean and remain submerged for long periods.

Carnivora These 240 species of mammals fall into two broad groups, one doglike and one catlike. The first group contains canids (dogs, wolves, and foxes), bears, procyonids, and mustelids; the second, viverrids, hyenas, and felids (cats). Most are solitary animals that live on freshly killed prey.

All carnivores have powerful skulls and jaws, with sharp teeth of various types that aid the animals in tearing their prey to pieces. They also have well-developed brains that give them an advantage as predators. Each limb has four or five digits ending in claws. Most carnivores are terrestrial or climbing animals and may be diurnal or nocturnal.

Most like the ancestral carnivores are the modern mustelids, consisting of weasels, martens, badgers, otters, skunks, and others. This group is small in body size relative to the other carnivores. They have short legs, long tails, and long, low bodies. Viverrids are close relatives of the mustelids, but are more catlike in having retractable claws and other characteristics. They include mongooses, civets, and genets.

The panda was once thought to be a close relative of raccoons, but recent studies have shown it to be more closely related to bears. Bears are quite doglike and are omnivorous. They are found both in Eurasia and the Americas, but not in Africa.

Proboscidea Elephants today live in two areas, Africa and southern Asia. In each area is one distinct species; these two species are all that remain of hundreds of extinct forms. Some of the mammoths lasted long enough to be hunted by early humans.

The two species appear quite different in the size of their ears and in the trunk or proboscis, a food-gathering organ formed by the nose and upper lip. The ears are used for cooling as well as hearing: when overheated, an elephant flaps its ears. Elephants have no canine teeth, and the single pair of upper incisors have been modified as huge tusks. They have huge legs that enable them to run fast but not jump.

Unlike most mammals, elephants have a matriarchal social structure. Related females live in herds that are headed by the oldest member, whereas males are solitary or live in small groups. Babies are cared for and guided for several years, either by the mother or by others in the herd.

Sirenia The sirenians (dugongs and manatees) are adapted to aquatic life almost as well as the cetaceans. Like whales, they have lost their hind limbs and live in the water for their entire life. However, their evolutionary origin is different; sirenians are descended from herbivorous ungulates, and cetaceans from carnivores. Sirenians eat aquatic plants.

These large mammals have small brains and are unintelligent. Lacking defense against such predators as sharks and crocodiles, they are confined to shallow coastal areas and estuaries where the predators seldom go. They eat marine algae and plants.

The sirenians (sometimes called sea cows) are named for the sirens of ancient myths because they nurse their young in a manner that resembles a human mother nursing a baby, reminding sailors of mermaids.

Perissodactyla The odd-toed ungulates include horses, zebras, tapirs, and rhinoceroses. All of them have long legs that end in blunt hooves. They have heavy cheek teeth adapted for grinding and elongated skulls. Their stomachs are simple. The axis of the foot passes through the middle digit in all odd-toed ungulates.

Once a large, diverse group, the perissodactyls have been reduced to a mere 17 species. Most became extinct in the Eocene or the Oligocene.

Artiodactyla The even-toed ungulates include 220 living species in several groups: pigs and hogs, peccaries, hippopotamuses, camels and llamas, chevrotains, deer, giraffes and okapi, pronghorns, and bovids (antelope, gazelles, goats, sheep, and cattle). Nearly all of the animals domesticated by humans are found in this order. They are natives on every continent except Australia and Antarctica.

In even-toed ungulates, the axis of the foot passes between the third and fourth digits. This is the chief characteristic that distinguishes them from odd-toed ungulates. In addition, they have stomachs with two to four chambers. In many species there is a regurgitation system allowing repeated chewing of the same food over a long period of time. The animals also have horns or antlers, a feature not found in the odd-toed ungulates.

NOTES

1. D. J. Boorstin, *The Discoverers* (New York: Random House, 1983), p. 421.

2. E. Malmeström and A. H. Uggla, *Vita Caroli Linnaei. Carl von Linnaeus självbiografier* (Stockholm: Almquist & Wiksell, 1957); quoted in Carl Linnaeus article in Swedish Museum of Natural History Web site, www.nrm.se/fbo/hist/linnaeus/linnaeus.html.en.

3. Charles Darwin, *The Origin of Species* (New York: Random House, 1950), pp. 373–74.

4. E. DeLong, "Archaeal Means and Extremes," *Science* 280 (April 24, 1998): 542–43.

5. Carl R. Woese, "On the Evolution of Cells," *Proceedings of the National Academy of Sciences* 99 (June 18, 2002): 8742–47.

6. Lynn Margulis and K. V. Schwartz, *Five Kingdoms: An Illustrated Guide to the Phyla of Life on Earth,* 3rd ed. (New York: Freeman, 1998).

7. Carl Zimmer, "All for One and One for All," *Natural History* (February 2002): 34.

8. Daniel S. Heckman et al., "Molecular Evidence for the Early Colonization of Land by Fungi and Plants," *Science* 293 (August 10, 2001): 1129–33.

9. "Plant and Insect Parasitic Nematodes," University of Nebraska–Lincoln Nematology Web site, http://nematode.unl.edu.

10. "The Life in Soil," Food and Agriculture Organization Web site, www.fao.org/ag/magazine.

11. "Understanding Rabbits and How to Control Them," Environment Southland (New Zealand) Web site, www.envirosouth.govt.nz.

6

Humans in the Biosphere

Though human beings are subject to the same limits that the environment imposes on all organisms, we are a unique species. Characteristics such as large brains and tool-using abilities, though shared to some extent by a few other animal groups, give us many selective advantages over other animals. Walking on two legs (bipedalism) instead of four freed early human hands for other purposes, such as using tools. Stereoscopic vision, which is possible only when the eyes are on the front of the head, is another human characteristic that has helped make it possible for us to use tools.

Our tool-using ability has been aided by movable fingers and opposable thumbs. Early hominids must have used stones and sticks from their surroundings as weapons and as food-getting tools. At some point, they also began making tools—breaking stones to create sharp-edged tools, scraping sticks to give them pointed ends, and so on. These tools extended the capabilities of legs, arms, and teeth. Today we extend the effectiveness of our eyes, ears, and other organs when we use microscopes, television, and computers. Even airplanes and rocket ships are elaborate tools that greatly extend the traveling abilities of our legs.

Domesticating plants and animals to provide a reliable food supply also helped early humans extend their own abilities, as did using energy (other than that in food) to perform work. Modern humans still benefit from these activities.

As early humans learned to domesticate some animals and to nurture useful plants, agriculture and medicine were born. In developing and testing tools for those crafts, workers became scientists. They learned to gather data, form hypotheses about the Earth, and carry out experiments.

But in spite of the obvious importance of tool-using, the capability that most sets us apart from other species is that of creating and using symbols, including language. The use of language is accompanied by a consciousness of self; to a slight extent, these characteristics are shared with some of the apes, but they certainly are distinctively human traits that must have furthered our cultural evolution. When early humans acquired speech, they could communicate with each other, begin to form alliances, offer praise or insults. Our ability to cooperate or compete with each other was greatly enhanced by language. Later, when we wrote down instructions and ideas, our power increased even more.

It is not yet known exactly what first made it possible for humans to speak. Speaking requires a combination of features in the brain, facial muscles, and respiratory apparatus. Some biologists have suggested that the Neanderthals' pharynx, for instance, was shaped differently than it is in modern humans.

Recently, geneticist Anthony Monaco and other researchers at Oxford University showed that the human version of a gene connected with speech, *FOXP2*, is no more than 200,000 years old. That was around the time that anatomically modern humans appeared. Mutations in the *FOXP2* gene cause various deficiencies in a person's speech and language, so it may influence the ability to make the mouth and facial movements needed for speech. Or language ability may already have evolved by the time the gene appeared. In either case, having *FOXP2* would have been an adaptation giving humans an additional selective advantage.[1]

However it evolved, language has set us apart and given us a modifying influence over Earth that is far beyond that of other organisms. Using the tools of language, science, and technology, early humans were able to turn large parts of the environment to their own purposes: They learned how to control fire and to cook food; the products of agriculture made it possible to feed larger groups of people; and even primitive medicine helped lengthen life, increase health, and make childbearing more successful. Early populations expanded as a result of all these advantages.

HUMAN POPULATION GROWTH

Anthropologists are still unsure about when and where modern humans first appeared. Some believe that *Homo sapiens* evolved independently in several places around the globe. But recent research on mitochondrial DNA and on human fossils supports the hypothesis that modern humans appeared in one location in sub-Saharan Africa and spread from there, replacing Neanderthals and other groups as they went.

Probably, according to what is called the Out of Africa hypothesis, modern humans are all descended from a single ancestral African group of humans. That group lived about 170,000 years ago, and their descendants continued to live on that continent until around 50,000 years ago. Then they began emigrating, eventually spreading around most of the planet.

One of the major events in human history, the Agricultural Revolution, occurred between about 10,000 and 7,000 years ago. Until then, early human populations survived by hunting and gathering food, a restrictive way of life that kept their numbers small (probably fewer than 10 million). But when they domesticated wild varieties of rice, wheat, and other food plants, people provided themselves with a reliable supply of food. They could stay in one area by farming, rather than hunting and gathering their food. Eventually stable groups of people could build villages, teach their children, and develop a division of labor. The Agricultural Revolution, along with the domestication of animals that were used for food or for helping with farm work, led to civilization.

When agriculture made it possible for them to live a more settled existence, human populations grew larger, and probably reached about 300 million by 1 A.D. Human population size generally continued to increase gradually through the following centuries, though it was held in check by disease, wars, and environmental factors. The most glaring exception to that steady rise was caused by a series of epidemics (the Black Death, possibly bubonic plague) beginning in the fourteenth century. In two years alone, the first epidemic spread across Europe and wiped out about one-third to one-half of the European population.[2]

Because population growth is geometric ($1 \rightarrow 2 \rightarrow 4 \rightarrow 8$) rather than arithmetic ($1 \rightarrow 2 \rightarrow 3 \rightarrow 4$), it follows a typical kind of growth curve (see Chapter 1). This begins with a lag period of slow growth. It ends, however, with a nearly vertical rise. The rise in human population size since about 1700 has been extremely steep.

During the past century, biologists and others have become increasingly concerned about the implications of this rapid growth. What is the Earth's **carrying capacity** for humans? At our current population of more than 6 billion, we are already facing crowding in cities and suburbs, water shortages, epidemics, and other difficulties even in the more developed countries. In less developed countries, the effects of overpopulation, such as famine, are far more serious. If the human population continues to grow at the present it may reach 11 billion in the twenty-first century. Will life be even tolerable for most people if that happens?

The story of one disease illustrates just one danger of crowded conditions. In 1918, as World War I was drawing to a close, a terrible epidemic raced around the Earth. Half a million died in the United States alone, 25 million in the world. In some countries half the population was lost to the disease. It was a pandemic (worldwide epidemic) of influenza. The newly mutated virus causing the epidemic appeared in France during the war and spread rapidly through the troops. Later it went to Spain and other parts of Europe, then spread to other areas, including the United States. Even today we fear the effects of the flu on the elderly and those with weakened immune systems, but the 1918 epidemic

attacked young and healthy people also. Lung damage, fevers, and secondary bacterial infections killed them.

Contagious diseases can spread like wildfire through crowded populations. With the flu, person-to-person contact makes it possible. With some other diseases, such as cholera, the virus enters drinking water from untreated sewage. In some cases (such as bubonic plague), fleas on rodents carry the bacteria, and a rise in the rodent population can lead to its transmission to humans. Once infected, a person may contract pneumonic plague if the bacteria spread to the lungs. They can spread to other people directly in coughs or sneezes. However it is spread, though, a crowded human population makes diffusion of the disease easier.

A huge human population is a danger to itself because of crowding, epidemics, and competition for resources. Wars—a major cause of loss of human lives—are much more likely when nations are competing for energy supplies of other resources. Overpopulation also endangers the rest of the biosphere. Though exceeding Earth's carrying capacity will inevitably lead to a population decrease, as discussed in Chapter 1, humans may be able to keep their population size in check voluntarily. The world population growth rate is slowing, owing to family planning in some countries and disease in others. Although the population increased by 64% between 1950 and 1975, it increased by 48% between 1975 and 2000.

HUMAN EFFECTS ON NATURAL RESOURCES

As our planet becomes more crowded, we are becoming very aware of diminishing resources. Some, such as fossil fuels, are being chemically changed to unusable forms; others, such as air and water, are

APPLICATION TO EVERYDAY LIFE FEATURE: DEBATING GMOs

In recent years the idea of genetically modified organisms (GMOs) has generated much controversy. Many people have been afraid to eat foods that have been genetically modified, thinking they may be harmful and even calling them "Frankenstein foods." GMOs are strictly labeled and controlled in much of Europe and Asia. In China, farmers are not allowed to grow genetically altered crops because they may not be able to compete in the skittish world food market (Kahn 2002).

That fear appears to be unjustified. Tests of corn, soybeans, and other foods have shown them to be harmless to humans and to have great advantages for farmers. Some GMOs are more resistant to insects than the unmodified organisms are, and some have more attractive colors or shapes. All of these attributes can increase the appeal of foods to consumers.

If GMOs used as foods pose no danger to humans and to have these advantages, why is there still resistance to them? It is partly because the possible genetic ramifications are still not fully understood. One possible problem has to do with allergies. Because altered DNA will produce altered proteins, and proteins can act as **allergens** in some people, some GMOs might cause allergic reactions. There have been a few cases of people who have eaten tacos and other corn products containing GMOs becoming ill afterward, but no link between the GMO and the illness has yet been established.

Another reason many people want to go slowly in adopting the use of GMOs is that the ecological consequences of introducing altered seeds into ecosystems might be harmful. For instance, in lab experiments a gene (from *Bacillus thuringiensis* bacteria) from a genetically engineered sunflower can pass to wild sunflowers, increasing their resistance to insects. That may seem harmless enough, but suppose the wild plant is a weed that is ordinarily held in check by insects. The weed can then invade croplands or natural areas where it is unwelcome.

One genetically engineered plant is Starlink corn, which was approved for use as a livestock feed only. Starlink corn also contains a gene from *B. thuringiensis,* and for that reason is called a Bt crop. The gene produces a protein that protects the corn against the European corn borer. That protein was a potential allergen and was thus banned for use in human food, though there was no evidence of its actually causing allergies in humans.

Though less than 1% of the U.S. corn crop was planted with Starlink in 2000, U.S. Department of Agriculture inspectors have found Starlink corn in 10% of grain samples from various parts of the country, showing it has already spread rapidly. Although the government is trying to control the spread by recalling seeds that may be contaminated, it may never be possible to remove it entirely from corn sold to humans.

According to the environmental group Friends of the Earth, nearly two-thirds of the commercially grown GMO crops (such as corn, soy, and canola) in the United States have been modified to tolerate herbicides or weedkillers. The original idea in modifying plants in that way was that fewer of those poisonous substances would be needed, and the environment would be protected. However, the result has been that many farmers now use two to five times the amount of herbicides as they once did, killing weeds without any worry about damaging their crops.

Another danger of Bt crops is similar to the overuse of antibiotics in medicine. Bt toxin has been used for years by organic farmers as one component of an integrated pest management scheme. It can be sprayed where needed to kill specific pests and is nontoxic to mammals. Now plants that produce the Bt insecticide themselves will select for insects that develop resistance to it. When many insects become resistant to Bt toxins, Bt sprays will be useless.

CONTINUED

Animals as well as plants may eventually become GMOs, though research on that area has not advanced far. According to the National Research Council, the biggest risk posed by developing genetically modified animals is that they might alter the environment.

Source: Kahn, J. 2002. "The Science and Politics of Super Rice." *New York Times,* 22 October.

becoming less available because human activities have polluted them.

Air

We tend to take breathable air for granted until it is polluted. Air pollution comes from many sources: Volcanoes periodically spew lava into the air and cause natural air pollution. Forest fires resulting from lightning are another natural source, and even the methane from cattle intestines pollutes the air. Vehicles, fires, and industry may add oxides of nitrogen, sulfur, and carbon to the air. In addition to these gases, polluted air may contain particles of soot or ashes that can affect comfort or health.

According to the World Health Organization, up to 150 million people in the world suffer from asthma, a respiratory disease that constricts the passageways for air. The incidence has been rising in all countries. The correlation between air quality and asthma is not yet definite, but it is generally accepted that air pollution triggers attacks among patients rather than causing asthma.

Most scientists believe emissions of carbon dioxide (CO_2) and other **pollutants** also threaten to disrupt climate and ecosystems worldwide. These materials rise high in the atmosphere and cause more of the sun's energy to be trapped beneath clouds or reradiated to the Earth (the greenhouse effect).

Because of strong economic growth in the 1990s, industrial expansion rose and CO_2 emissions increased. U.S. emissions of CO_2 from burning fossil fuels rose 2.7% just in 2000, according to the Department of Energy's Energy Information Administration. Because CO_2 emissions make up more than 80% of U.S. greenhouse gas emissions, they are considered a good indicator of what to expect in the nation's total greenhouse gas emission levels. In the future the Earth's climate may change suddenly if large emissions of greenhouse gases occur.

By emitting pollutants from vehicles, industries, and even consumer items (such as hairspray or other aerosol items) we humans have also affected the layer of ozone that shields us from the sun's ultraviolet radiation. Thinning of the ozone layer was first observed in the 1950s over the Antarctic and since 1999 has shown up over the Arctic as well. The hole over the Antarctic was found in 2002 to have shrunk somewhat since 2001, but scientists studying it thought it was probably a temporary improvement.[3]

The greenhouse effect and thinning of the ozone layer both contribute to global warming, which may have drastic changes on sea level and on ecosystems. If the Earth's temperature rises even a few degrees, polar ice caps may melt, causing a rise in sea level. Coastal areas in all countries may be submerged or become wetlands. New wetlands can provide new habitats for a variety of plants and animals that are adapted to those conditions. Farther inland, rangelands may become deserts.

Midwestern severe winters may become much milder, and hot summers become even hotter.

As the climate changes, changes in ecosystems will follow. One early indication of such effects is the death of amphibian eggs in the wild, according to a study by Joseph Kiesecker, a biology professor at Penn State University. Scientists have been trying to find the reasons for declines in amphibian populations around the world for more than 20 years and have blamed the declines on a variety of causes. Kiesecker says his team has shown global warming causes changes in rainfall patterns. This leads to stress in moisture-sensitive amphibians and makes them susceptible to various pathogens. The specific stresses and causes of death differ with conditions in the animals' habitats.[4]

Surprisingly, some atmospheric pollutants may be having favorable effects. In a study reported in 2001, some scientists found that the particles of pollutants in clouds cause condensation around themselves, resulting in water droplets that reflect the sun's rays away from Earth and leading to cooling of the areas beneath the clouds.[5] This condensation is similar to the natural condensation around particles in clouds that is part of the water cycle shown in Chapter 1.

Thanks to provisions of the Clean Air Act, which was passed about 30 years ago, the overall quality of air in the United States has increased. This has been an encouraging development, showing that human-caused environmental degradation is not inevitable.

When pollutants in the air are washed out by rain or other precipitation, the air itself becomes fresh and clean. The pollutants do not go away, however; they enter rivers and other bodies of water, adding to water pollution. Or the pollutants react with water in the air to become **acid precipitation.** Sulfur dioxide and oxides of nitrogen are the usual pollutants that form acid precipitation.

Water

Off the coasts of Louisiana and Texas in the Gulf of Mexico is a vast dead zone— a huge oxygen-poor zone caused by fertilizer runoff from farms and lawns. Already 22,000 square kilometers in area, it is growing larger. Government scientists are studying ways of restoring the area to save the fisheries there.

Fertilizer (which contains nitrogen, potassium, and phosphorus) is a major pollutant in waterways. Either manure or artificial fertilizers can damage aquatic ecosystems. Phosphorus and other nutrients in fertilizers cause algal blooms that can cover most of the water surface; this process, called eutrophication, nearly destroyed Lake Erie in the 1960s.

The effect of small amounts of fertilizer almost seems positive; aquatic plants are stimulated to grow by the nutrients. However, if the plants are not eaten by birds or fish, they eventually die. The decomposition process uses oxygen and releases CO_2 and other products into the water. In time, if many plants die the level of O_2 may fall so far that animals die, adding further to the decay and pollution in the waterway.

Fertilizer is only one of the water pollutants resulting from human activity. Sewage, industrial chemicals, paints, oils, and trash all have been found in waterways, causing ugliness and possible danger to aquatic ecosystems. Some of the pollutants in water are chemically similar enough to human hormones that if they enter human bodies, they fit into receptors on cell surfaces and bring about or block the effects that the hormones would have normally.

Many pollutants act indirectly on humans by moving through food chains.

Mercury is one example; it is not metabolized by animals, as some pollutants are, but is concentrated in their livers. At each step in the food chain, a higher level of mercury is passed on.

Soon after the Gold Rush brought hopeful miners from all over the world to California, the early panning for gold technique gave way to hydraulic mining. Not only did that method damage the hillsides that were flooded to expose gold ore, but today hydraulic mining has even left a legacy of poisoned fish in the Sierra Nevada Mountains. Mercury was used to separate gold from the ore by forming an amalgam with it, and mercury ended up in the waterways. Bass and catfish now have high levels of mercury, and mercury poses a danger to humans eating those fish; it can lead to nerve damage and developmental disorders. Other fish in the area, such as sunfish and trout, contain lesser amounts of mercury but are being monitored by the state of California. Ocean fish, especially swordfish and tuna, contain high levels of mercury and are dangerous to health, especially for developing fetuses.

Soil

Drought has afflicted many areas of the contiguous United States in the past several years. Farmers across the country have watched their crops dry up in the hot sun. July 2002 was the fifth-warmest July on record. In the West, several states were declared agricultural disaster areas by the U.S. Department of Agriculture, and more than 4 million acres of forest burned in wildfires.

Surprisingly, the drought was not the result of low precipitation: 6.8 cm (2.68 inches) of precipitation fell in September 2002, 0.58 cm (0.23 inches) more than the 1895–2002 average. However, the average September temperature was 19.8°C (67.7°F), which made it the seventh-warmest September in that period and led to much evaporation.

Drought is a continuing problem, affecting some part of the United States almost every year. Several government agencies, in cooperation with the University of Nebraska–Lincoln, provide useful animated drought-monitor U.S. maps on the Web (visit www.drought.unl.edu/dm/monitor.html). They are frequently brought up to date. The maps show normal areas of the country as well as those ranging from abnormally dry to drought–exceptional.

In the Dust Bowl years of the 1930s, drought helped lead to the loss of valuable topsoil from the center of the country. Since World War I, western land suitable only for grazing had been used for growing wheat, which failed to hold the soil in place as well as the native grasses did. Dry soil without plant roots to anchor it is likely to blow away in the wind. Erosion of the topsoil, by wind and by flood waters, was a major national problem in the 1930s; as the soil washed into rivers and finally the sea, the livelihood of much of the nation vanished with it. Though erosion does not destroy the soil but just moves it to another place, farmland without topsoil is useless. Its natural restoration can take years. If global warming leads to more evaporation, drought effects will become even more severe in the coming years.

Although drought can lead to erosion of large quantities of topsoil if much natural vegetation has been removed, human activities can also affect the quality of the soil. Continuous culture of tobacco, for instance, can cause serious degradation of the soil. As nitrogen and other compounds in the soil are taken up by plants that are harvested, the compounds must be replaced if the soil is to remain fertile.

Soil is a renewable resource, but building fertile soil takes time. Conserving soil is much faster and less expensive than restoring it after important compounds are lost.

Minerals

The elements in minerals—naturally occurring inorganic substances having a definite chemical composition—cycle through the biosphere, just as elements in living organisms do (see Chapter 1). However, minerals tend to be formed more slowly and last much longer than organisms do, so the elements may remain in mineral "sinks" for long periods. For example, the phosphorus that is needed by all living things may be tied up for centuries as the phosphate in rock. If the rock is mined for use as fertilizer, it dissolves in soil water and is taken up by plant roots in the fertilized soil. It then becomes part of the plants' nucleic acids and is passed on through the food web. It can return to rock only after a long period, because phosphorus must be washed into waterways and be deposited in sediment that is finally turned to rock. When we remove minerals from the earth, we are removing them virtually forever.

Many minerals are needed in human foods or for industrial purposes (see Chapter 4). If we need larger quantities of a mineral than we can obtain from the living part of the elements' cycles, we must conserve the mineral stored in the nonliving part.

A major storage place in the phosphorus cycle is phosphate rock, which contains the inorganic ion phosphate (PO_4^{3-}). The rock may be used in fertilizer for agriculture, where it breaks down and enters soil water. Plants take up the phosphate through their roots and incorporate it into their DNA; when they are eaten by animals, the phosphate moves up the food chain. Because phosphorus is not given off in respiration, it does not return to the soil or water unless it is in the form of compost or sewage sludge. It may end up in landfills.

Human activity can have a major effect on the phosphorus cycle. For example, when tropical rain forests are cut down, the phosphorus stored in the trees and shallow leaf litter is soon washed away by rains. (In these forests, the soil has scarcely any reserves of nutrients.) It eventually washes into waterways, where it causes pollution. At the same time, coffee or other crops planted where the forest stood must be fertilized with phosphorus that usually comes from phosphate rock. If the fertilizer is manure or sewage sludge, the natural cycle can be restored. When coffee is shade-grown, the natural rain forest remains and prevents erosion by rain.

Energy

Human use of electrical and chemical energy is immense, especially in industrialized countries, so there is continual pressure to extract **fossil fuels** wherever possible. The current administration in the United States has strongly urged opening the coastal plain of the Arctic National Wildlife Refuge (ANWR) to drilling for oil, in an effort to lessen dependence on foreign sources. This has led to strong opposition from some environmentalists who think the environmental costs would outweigh the benefits of obtaining new oil and natural gas supplies. They claim that the ANWR has only a six-month supply of oil and that the area that would be drilled is where caribou from Canada's Porcupine River region migrate in the spring and give birth to their calves. According to the U.S. Fish and Wildlife Service, oil development in the coastal plain would cause the Porcupine River caribou population to decline or move to less favorable habitats with less food and many more predators. Other animals, too, call the coastal plain home: Nearly 200 species of wildlife live there.

Drilling proponents counter those arguments by claiming that caribou remain there only briefly. They also maintain that oil production would be spread out over many years and at peak output could cut imports by 1 million barrels a day. A Web site maintained by the Arctic Power Company (www.anwr.org) speaks of the coastal plain as an "arctic desert," an excellent environment for annoying insects, and a very promising source of new oil. The company lists several reasons for supporting development.

1. Only 8% of the ANWR would be considered for exploration; the remaining 17.5 million acres would remain permanently closed to development.

2. Federal revenues would be enhanced by billions of dollars from bonus bids, lease rentals, royalties, and taxes.

3. Between 250,000 and 735,000 jobs are estimated to be created by development of the coastal plain.

4. Between 1980 and 1994, North Slope oil field development and production activity contributed over $50 billion to the nation's economy.

5. U.S. Department of Interior estimates range from 9 to 16 billion barrels of recoverable oil from the coastal plain.

6. The North Slope oil fields currently provide the United States with nearly 25% of its domestic production of oil, and since 1988 this production has been on the decline.

7. The United States imports over 55% of the nation's needed petroleum, which cost more than $55.1 billion a year.

8. Oil and gas development and wildlife are successfully coexisting in Alaska's arctic. For example, the Central Arctic caribou herd at Prudhoe Bay has grown from 3,000 to as high as 23,400 during the last 20 years of operation.

9. Advanced technology has greatly reduced the footprint of arctic oil development.

10. More than 75% of Alaskans favor exploration and production in ANWR. The Inupiat Eskimos who live in and near ANWR support on-shore oil development on the coastal plain.

The ANWR issue is intensely political, and there are powerful arguments for both conservation and development. As the United States and other more developed countries have used greater amounts of fossil fuels, the possible sources of these fuels have become scarcer. At one time, the idea of destroying a wildlife refuge for the purpose of oil drilling would have seemed inconceivable; today, with diminishing resources and the threat of supplies from the Middle East being interrupted, every source must be weighed.

Fossil Fuels

Fossil fuels were formed during the Carboniferous period, when the plants and animals in swamps died and were eventually compressed and changed chemically into coal, oil, and gas. Theoretically they could be replaced in time, but it would take hundreds of millions of years.

OIL

The hydrocarbon compounds in crude oil are a closely related mixture of substances such as gasoline, lubricating oil, and paraffin. They can be separated by **fractional distillation,** as each one vaporizes or condenses at one specific temperature. Because oil was formed by the decay of plants and animals, crude oil can also be contaminated by spores, bits of skeletal material, and other plant and animal remains. Plankton was probably the origin of most oil.

Oil has been an important product for humans for thousands of years, providing medicines, lubricants, and fuel for lamps.

For a long time, whale oil was a major source, but as the whaling industry expanded in the nineteenth century and thousands of whales were killed, that source lessened.

Around the same time, prospectors began exploring for oil fields and found some gigantic ones. Although the United States and a few other countries have some large oil fields, about two-thirds of the world's oil is in the Persian Gulf region. The United States, on the other hand, is the largest consumer of oil in the world, representing more than one-fourth of the total world demand for oil. If world use of oil continues at the present rate, according to present supply and use rates scarcely any oil will be left by the end of the twenty-first century. (The total amount of oil remaining is unknown.) Obviously, some changes in oil sources or usage must take place if we are to maintain anything like our current automobile-dependent way of life.

NATURAL GAS

Most petroleum is found as either a liquid (crude oil) or a gas (natural gas). Natural gas is a mixture of hydrocarbons with very low boiling points. In nature, natural gas may be mixed with crude oil that is under pressure or may form a cap above pools of oil. It may also be found alone in spaces in rock.

At one time, before the 1960s, natural gas was used little except for home heating; it was considered a rather useless by-product of petroleum refining and was always removed from petroleum by flaring during the refining process. More recently, as oil supplies have dwindled, natural gas itself has become an important fuel, and some refineries recapture the gas that would otherwise be burned. Because it can be transported in pipelines rather than in trucks and burns cleanly to produce only CO_2 and water, it has advantages in addition to its availability.

There are huge reserves of gas in Russia, North America, and the Middle East, as well as smaller reserves elsewhere. As gas usage increases, these are likely to shrink rapidly. As we use up the remaining reserves of oil and gas, we may turn to another fossil fuel, coal.

COAL

Coal was formed during the Carboniferous period by the compaction of dead plants. These first formed soft peat deposits that later changed to harder forms of coal, such as bituminous and the less abundant anthracite.

Coal has been in use for at least 2,000 years. People used coal mainly for home heating until the Industrial Revolution in the eighteenth century, when the steam engine and other new technologies created a great demand for coal. Even with industrial use, there is still a great amount of coal available, and it can be converted to clean fuels that avoid some of the soot and other environmental problems caused by using solid coal. However, when coal itself is burned it contributes greatly to the greenhouse effect by sending particulate matter and CO_2 into the atmosphere.

The largest reserves of coal in the world are in Russia, Ukraine, Kazakhstan, the United States, and China. In the United States, most coal is found in the Midwest and in the Appalachian Mountains. Coal mining in the United States has often led to scarring of the landscape, though recently there have been efforts to restore areas that have been damaged by mining.

Coal gas, or coal that has been converted to gas, has been used for centuries, but until recently its quality was poorer than that of natural gas. Now a high-quality coal gas, or synthetic natural gas is being produced that can be piped through existing natural gas pipelines. The largest source of synthetic gas in the world is the Great Plains Synfuels Plant in Beulah, North Dakota. Unlike other fossil fuels, coal is still available in large amounts. At the present rate of use, it can provide energy for hundreds of years.

Other Fuels

Oil and gas have met most of our energy needs well until now, but even if conserved they are unlikely to last long enough to provide fuel for the growing world population. Alternatives must be found.

NUCLEAR

The fossil fuels pose two major problems: diminishing supplies and pollution. To some extent, using nuclear energy would solve these problems.

In learning how to benefit from the fission of atoms, humans entered a new era in which they were no longer limited to burning fuels that had accumulated naturally. Uranium ore, consisting almost entirely of U-238, can be enriched to increase the percentage of U-235, the form of uranium that will fission and release energy.

Using energy from nuclear fission will not be possible indefinitely: The reserves of uranium ore will outlast those of fossil fuels, and by 2100 even uranium may be in short supply. In addition, although nuclear energy does not produce the soot and other air pollutants associated with fossil fuels, it does cause thermal pollution because water is used to cool the nuclear reactors. Radioactive wastes are produced as well, causing a major waste-disposal problem.

Breeder reactors have been suggested as a solution to the problem of a shortage of uranium. In these reactors, the plutonium produced in fission is reused to produce new energy. If a safe breeder reactor can be perfected, the diminishing-fuel problem can be at least postponed. However, a breeder reactor will pose the same issues of waste disposal and pollution that current nuclear plants pose. In addition, the fissionable materials they produce could conceivably be used by terrorists to make nuclear weapons. Nevertheless, some countries are pursuing the breeder reactor option. For example, India, which has little uranium, is planning to build five fast breeder reactors by 2020.[6] Russia has one fast breeder reactor, but France shut down its own in 1998. In the United States, though interest in nuclear power has waned, research is continuing. It may be possible, for instance, to find an economical and practical use for the waste, such as in home energy systems.

Nuclear fission power is based on splitting atoms of uranium **isotopes,** but the nuclear fusion reaction is based on combining atoms of hydrogen isotopes (deuterium and tritium). Deuterium and tritium are both either abundant in nature or products of abundant natural elements. In fact, the deuterium in sea water could produce all the Earth's needed energy for the next billion years.

Although fusion offers the promise of abundant, cheap fuel, it also poses problems, and using energy from nuclear fusion is a goal that has so far eluded energy producers. Tritium atoms are radioactive and would pose a radiation danger if they escaped into the atmosphere. Also, bringing about the fusion reaction requires enormous amounts of heat and pressure. At this time, no acceptable fusion reactor is available. An enormous and expensive multinational project, the ITER fusion reactor, is under way in Ontario. (ITER is not an abbreviation or acronym but the Latin word for "the way.") According to environmentalists opposing nuclear power, the ITER project's directors admit that they are at least 50 years away from creating a sustained fusion reaction.[7]

GEOTHERMAL

Geothermal power is used in some areas, such as Iceland and parts of northern California, where naturally radioactive materials in the Earth's core heat rocks within the **crust.** The rocks then convert ground water coming in contact with them into steam. The steam can be piped to the Earth's surface and sent through a turbine generator for producing electricity.

Although this might sound like an inexhaustible source of power, the underground water supply may be limited. The steam also may dissolve sulfur and other mineral salts, leading later to air and water pollution or to problems with land waste disposal.

SOLAR

Solar power certainly poses no problems of supply—the sun is always out there sending free power. However, on cloudy days, it is not available. Thus solar energy is useful mostly in areas where there are many sunny days each year.

Humans have had no effect on this resource, either on the supply or on producing pollution from it. When and where solar power can be used, it is extremely effective in lowering utility bills and lessening pollution. Though not a major source of energy now, it is likely to become used more as fossil fuels are used up.

WIND AND WATER

Using water or air for generating electricity has the important advantages of creating no pollution and starting with free, abundant materials. Hydropower has been used for years, with the potential energy of the water behind a dam driving large turbines that turn and generate electricity. Though very useful under the right condi-

tions, it cannot be used where there is no rapidly flowing river or during a drought. In addition, dams have interfered with salmon migrations and caused other environmental problems. As a result, few dams are constructed today.

The force of tidal water also has been harnessed as a source of electrical power but can be used in only the few places where the vertical distance between high and low tides is great and where a dam can be constructed. The electricity is generated according to the tidal cycle, which is not always convenient for human purposes.

Windmills also can be erratic, generating electricity only when the wind is blowing. The large windmills that produce considerable amounts of electricity are noisy and may kill eagles and other birds that glide on wind currents. Windmills are used successfully in some windy places, such as Denmark and the passes between mountains in California, but they generate opposition from neighbors as well as electricity.

HYDROGEN POWER

Hydrogen is everywhere around us, especially in water (H_2O). Its use as a power source is currently being promoted by futurists who think hydrogen fuel cells are the answer to the fuel supply and pollution problems of cars and other vehicles. They are already used in experimental buses in Chicago and in a prototype automobile by General Motors.

Like solar power, hydrogen would have the advantage of an abundant natural supply. Also like solar, it would cause no pollution; hydrogen fuel cells produce only water.

SCIENCE AS A HUMAN ENDEAVOR

Science is not only technology and laboratory and fieldwork; sometimes it is intertwined with creative, even emotional work. Nor is it free from values.

Parents, peers, teachers, and religion all affect our values. These influences may differ widely, so that what seems good in some cultures may seem bad in others. Thus our values help give us a subjective way of making decisions.

Logic and scientific methods and processes, however, provide objective evidence on which to base decisions. Throughout history, humans have made observations, made and tested hypotheses, and reached tentative conclusions. Those conclusions usually have been revised or rejected in time, to be followed by new ones. Like our use of language, our scientific search for knowledge is unique to humans. No other animal is a skeptic.

Science has led to a large body of knowledge, but science is more than facts, more than broad concepts and theories. Science is the use of logical processes in a search for knowledge that never ends. In a series of monographs about biology education, late evolutionary biologist John A. Moore (1915–2002) called science "a way of knowing."

There are other ways of knowing, Moore said, that may lead to truth. In searching for truth we may find many answers in religion, philosophy, and art. For some questions, science can provide no answers at all. However, in searching for factual information about the natural world and in applying knowledge to technology, we need to use science. Science has certain characteristics that set it apart from other ways of knowing: Scientific explanations are based on making critical observations about nature and on conducting experiments with natural phenomena. Because of this dependence on observation, scientific explanations lead to testable predictions about nature. These predictions are tested in

the open, and the results are both made public and subject to criticism by other scientists. All results are tentative, and subject to change as new evidence is found.

Nowhere are logic and scientific evidence more crucial than in analyzing past history and current trends to predict and influence the future. The survival of the planet may depend on what politicians and other people do today to manage ecosystems for future generations.

THE FUTURE OF THE BIOSPHERE

Near the close of the nineteenth century, there was a sudden interest in wilderness and forest conservation brought about by the census. In 1880 large wilderness areas of the United States were "uninhabited"—that is, had a population of fewer than two persons per square mile. But only 10 years later, the census of 1890 showed that no such areas remained. That caused many Americans to look more carefully at the country that remained and to consider what would happen to the land in the future.

Early in their history, Americans' attitude toward wilderness had been fearful; from the mid-nineteenth century onward, however, that attitude changed.[8] It changed in two ways—toward wilderness preservation and toward forest conservation. To romanticists and to some naturalists (such as John Muir [1838–1914], who founded the Sierra Club), wilderness was to be preserved in its natural state, where men could feel themselves "part of wild Nature, kin to everything."

Muir did not create his wilderness philosophy; rather, he restated poet and naturalist Henry David Thoreau's (1817–1862) ideas of 50 years before. The philosophy that had been little understood in Thoreau's time found a receptive audience throughout the country by 1900, and Muir became a best-selling author who influenced millions of Americans.

In California, those who wanted to preserve wilderness focused their efforts on the redwoods. Muir and other early members of the Sierra Club began a continuing battle to save the redwood forests.

However, the John Muirs were outnumbered not only by private developers who wanted to reclaim water for irrigation and use forests for lumber, but also by the Gifford Pinchots. Pinchot (1865–1946) was the first chief of the U.S. Forest Service and the person who claimed to have coined the phrase "conservation of natural resources." Pinchot's approach to conservation was one of forest management rather than wilderness preservation. He had been trained in the European style of forest management techniques and was responsible for many forest preserves being saved by private developers. But Pinchot believed in managing resources wisely and responsibly, not necessarily in leaving them untouched.

That attitude clashed with that of Muir, who had previously been Pinchot's ally; when Pinchot planned the 1908 White House Governors' Conference on Conservation for President Theodore Roosevelt (1858–1919), he invited industrialist Andrew Carnegie (1835–1919) but not Muir.[9]

In the following years, industry became more important in American life, as two world wars demanded ships, airplanes, and weapons. Also, the focus of conservation shifted from forests to fields during the Great Depression years, when soil erosion led to dust storms that ruined many farmers. Technology provided the answers to many agricultural and medical issues, giving Americans antibiotics, antitoxins, vaccines, insecticides, and other products for controlling the natural environment. Thanks to stringent public health measures, tuberculosis and other contagious diseases declined.

Newly discovered vitamins were manufactured, promising a shortcut to preventing disease. Rivers were dammed for controlling floods and providing hydroelectric power.

In the 1950s, few people worried about the biosphere's future. The natural world seemed a stable place, aside from the nuclear bombs that threatened human and other life, and scarcely anyone used the term *environmentalist*. In everyday life, plastics were being used for furniture, dishes, and much else. Synthetic "wash and wear" fabrics became common.

To be sure, some early twentieth-century ecologists adhered to the importance of conservation. One was Wisconsin biologist Aldo Leopold (1887–1948). Leopold preached a land ethic based on preserving the natural environment; he said that "a thing is right when it tends to preserve the integrity, stability and beauty of the biotic community."[10] However, the general public did not yet share that concern.

Around 1960, biologists led by Rachel Carson (1907–1964) began sounding the general alarm, telling us that pesticides and other pollutants were traveling through food chains and poisoning the environment. Carson's book *Silent Spring* suddenly made much of the country aware of the danger.

Population issues became important at about the same time. Speaking at the University of California in Berkeley in 1962, ecologist Marston Bates (1906–1974) compared human population growth to the growth of a cancer:

The multiplication of human numbers certainly seems wild and uncontrolled. The present annual increase in human numbers on the planet is about 48 million individuals. Four million a month—the equivalent of the population of Chicago. And whatever one thinks of Chicago, a new one every month seems a little excessive. We seem to be doing all right at the moment; but if you could ask cancer cells, I suspect they would think they were doing fine. But when the organism dies, so do they; and for our own, selfish, practical, utilitarian reasons, I think we should be careful about how we influence the rest of the ecosystem.[11]

Forty years later, the human population is continuing to grow and influence other life on Earth. Thanks to Carson, Bates, and other far-seeing ecologists, however, we are at least more conscious of our actions today. National and international groups have sprung up, eager to control human population growth, diminish our use of resources, and lower the levels of pollutants we discharge into the environment.

Even within the ranks of environmentalists there are disagreements about how to bring about desired results. For instance, in the United States there is sometimes tension between the environmentalists who welcome immigrants and those who think new people simply add to our current population-related problems. Population biologist Garrett Hardin, for instance, has written eloquently that we need a "lifeboat ethic" based on the assumption that bringing more swimmers aboard a crowded boat will simply drown everyone in the boat.

Even more pronounced are the differences between environmentalists and such businesses as logging companies, oil companies, and others who depend on using energy and raw materials for making their products. These businesses may employ great numbers of people, and they usually have more influence on local economy and politics than environmentalists do.

Environmental issues, like those in modern genetics, are not just matters of liberalism or conservatism, Republican or Democrat. They are complicated questions requiring us to think carefully about costs, benefits, and ethical principles. For instance, the ban on DDT brought about largely by *Silent Spring* has allowed the spread of diseases carried by insects, such as malaria. Thus, attempting to solve one problem has created another.

New Approaches to Conservation and Controlling Growth

Some progress in controlling air pollution is occurring. In 1996, the U.S. National Oceanic and Atmospheric Administration (NOAA) reported that the ozone layer was beginning to recover, thanks to global efforts to control the pollutants that damage it. They even predicted that the Antarctic ozone hole will have closed in 2050. In 1997, 150 countries sent representatives to Kyoto, Japan, where they agreed to reduce worldwide emission of greenhouse gases 5.2% by 2010. (The United States did not sign the agreement, arguing that it called for controls on the more developed countries only and that the less developed countries are among the major polluters.)

The Clean Water Act of 1972 was passed in an attempt to reverse the damaging effects on water of decades of misuse. Like the Clean Air Act, it has made a difference: Waterways have improved as a result, but some pollutants remain. Landfills and other sites have been used for disposal of pollutants that would formerly have ended up in water.

The 2002 World Summit on Sustainable Development in Johannesburg, South Africa, brought together leaders from around the world. (Sustainable development, the general theme of the Johannesburg conference, refers to agricultural and other practices that can be continued indefinitely without damaging the environment or depleting it of energy or material resources.) There were some exceptions; the United States, for one, failed to send a high-level delegate. Although the overall goal was to promote sustainable environments, this summit emphasized human health and economic development more than previous meetings, such as the 1992 meeting in Rio de Janeiro, had.

The Johannesburg summit, more than earlier meetings, focused on issues of human resources, such as disease and poverty. Though environmental problems may affect humans, addressing pollution and resource depletion does not necessarily improve life for individual humans. Local politics may affect the distribution of food sent for relief, for example. Cultural practices of a group may lead to the spread of sexual and other diseases.

Partly because the United States was little involved in the Johannesburg meeting, few expected much from it. It did result in more than 100 governments agreeing to cooperate to reduce poverty and protect the environment. Delegates agreed to restore fisheries to maximum sustainable yields and establish a network of protected marine areas. In addition, they set goals—slowing biodiversity loss, substantially increasing use of renewable energy, and greatly increasing the number of people with access to sanitation. All the goals are supposed to be met by 2015 or earlier. Time will tell whether these goals are met; the nations represented are not bound by the plan of cooperation, and there are no specific plans for individual countries.

The Johannesburg emphasis on health was not lost on American scientists. An editorial in the prestigious journal *Science* stated that "a wide appreciation of the ecological significance of population health, viewed over decadal time, is vital to enrich the prevailing, usually superficial, shorter-sighted discourse on sustainable development."[12]

As the editorial pointed out, it is important to think of sustainability in reference to health, not just in reference to ecology and economy. Throughout history, environmental and human health have been intertwined, for good or ill: Humans have acquired infectious diseases from livestock and primates; exploration, war, and trade have contributed to epidemics; some agricultural practices have depleted the soil of micronutrients; urbanization and industrialization have both helped control some diseases and helped disperse others; and the spread of high-fat, sugary processed foods has contributed, along with the sedentary lifestyle made possible by automobiles, to obesity and related diseases. The environment can be changed in ways that benefit human population health or in ways that damage it. Any international group concerned about health needs to promote conserving an environment that sustains human health as well as the biosphere itself.

Maintaining Biodiversity

The diverse life on Earth has evolved over hundreds of millions of years, each species affecting the species surrounding it and being influenced by them. Disturbing any part of the web of life may have far-reaching results on the living and nonliving environments, so preserving biodiversity (the diversity of organisms and ecosystems in the natural world) is of major importance. Many conservation groups are involved in preserving biodiversity in the face of global urbanization and development.

Earth still has an enormous number of different species: for example, sociobiologist E. O. Wilson stated recently in a National Public Radio interview that it is common for just one tree in the Amazon rain forest to provide habitats for 50 different species of ants.[13] However, we are losing species of all kinds at the rate of about 1% every year, because of the expanding human population. According to a recent study, scientists now estimate that as many as 47% of the world's plants are endangered species.[14]

James Kirchner, a professor of Earth and planetary science at the University of California at Berkeley, carried out a study of species' appearances and extinctions based on a fossil database created by late University of Chicago paleontologist Jack Sepkowski (1949–1999). Kirchner cataloged all the species of marine animals fossilized during the past 530 million years. He learned that although species can die off rapidly, it may take millions of years for a new niche to develop and for a species to be created for it. The implications for humans are obvious; as Kirchner says, "There's not a lot of evolution going on in a parking lot."[15]

U.S. and international groups, such as the Nature Conservancy and Conservation International, have taken the approach of raising money to buy large tracts of land and protecting it from development or exploitation. Though expensive initially, this is an effective way of preventing long legal battles about how land can be used, property-owners' rights, and so on.

When large land or water areas can't be protected, habitats for native species may disappear or become polluted. Another threat to native species is competition from nonnative species that may be introduced deliberately or by accident. One example of that threat can be seen in Lake Erie: A botulism epidemic hit parts of the lake in summer 2002, killing more than 8,000 common loons and thousands of turbot, perch, and other fish. *Clostridium botulinum,* the bacteria causing botulism, produces a powerful nerve toxin. The animals had ingested mussels, which filter water to get food. Apparently the mussels had taken in the bacteria and concentrated the toxin in their bodies. When the mussels were eaten by fish and birds, those animals were poisoned.

C. botulinum are always in the environment in low amounts, but in 2002 the ecosystem was probably changed by invader species of zebra mussels and quagga mussels from eastern Europe. As the mussels filtered algae out of the water, the water was made clearer, allowing sunlight to penetrate the lower part of the lake and stimulate plant growth. As in eutrophication, the plants eventually died and decayed, using up oxygen. Because botulism bacteria thrive in low-oxygen environments, they expanded greatly and made the outbreak possible.[16]

Another nonnative species that has caused damage to the environment is the brown tree snake *(Boiga irregularis).* The snakes arrived on Guam in the early 1950s. (They probably were hidden on ships from the New Guinea area, about 800 miles away.) By the late 1960s, they had probably spread throughout the island. The snakes had evolved elsewhere, so they had no natural predators to control their population on Guam. Though pigs and monitor lizards eat snakes, they have not lowered the snake population in most areas of Guam, and today some forested areas in Guam may have as many as 13,000 snakes per square mile. The snakes feed on lizards, small mammals, chickens, and native forest birds and have nearly killed all the forest birds. Some of the native birds had lived nowhere else, so they are gone forever. Other species are now close to extinction. The snakes have also caused damage to the nonliving environment by crawling on electrical lines, causing power outages and damaging the lines.[17]

TOOLS: GLOBAL VIEWS OF EARTH

Ecosystems can now be studied from computer-equipped satellites far from Earth. A geographic information system (GIS) can assemble, manipulate, and display nearly any information in relation to a region's geography. For instance, data about vegetation or rainfall in various areas can be linked to aerial photographs of those areas. Any variable that can be located by latitude, elevation, and longitude, or even by ZIP code, can be fed into a GIS. The data can be added in map form or can be converted to a map.

NOAA has used GIS technology to map changes in the ozone hole over the Antarctic. You can see dramatic animations of the changes over time at many NOAA and Environmental Protection Agency Web sites (such as www.epa.gov/ozone/science/hole). The ozone hole data is in units of ozone in each location, but on GIS maps it appears as purple, red, burgundy, and gray areas that have appeared over Antarctica since fall 2000.

Biologists use collar transmitters and satellite receivers to track the migration routes of animals like caribou and polar bears. For one study, after manipulation by a GIS, the migration routes were shown by different colors on maps for a series of months. The maps of migration were then superimposed on maps of oil development plans to determine the potential effects of development on the animals.

Using GIS technology, utilities can plan where to install new water wells to avoid toxic sites that can be located and mapped. Several sets of data can be combined to form a site selection map. It shows areas that are undeveloped, are located outside the polluted areas, and lie above water-saturated sand and gravel by 12.2 m (40 feet) or more.

The maps are also useful for simulating the effects of altering environments. In one case, the National Forest Service was offered a land swap by a mining company seeking the development rights to a mineral deposit in Arizona. Using a GIS and varied digital maps, the U.S. Geological Survey and the Forest Service created dramatic maps of the area, showing how the terrain would be changed by mining.

NOAA carries out aerial surveys of areas such as the Channel Islands National Marine Sanctuary, a few miles off the coast of Santa Barbara, California. Data are collected and analyzed about the distributions of kelp, marine mammals, and other plants and animals. The resulting maps are used for educating the public, protecting the resources of the sanctuary, and monitoring ship and airplane traffic in the area.

Many of the predictive climate computer models have been criticized as not producing accurate results because the data on which they operate are unreliable. Though improvements may be needed, GIS technology seems like a promising tool for biologists.

Lessening the Impact of the Built Environment

Our buildings, roads, and other parts of the built environment have come to dominate more of the Earth, crowding out the plants and animals of the natural environment. Though this is inevitable with human population growth, some planners are coming up with new ways of lessening the negative effects. A new movement called green architecture, for instance, has focused on reusing natural materials in building, increasing natural ventilation, lowering energy use, recycling waste water, and similar strategies. Though sometimes more expensive at the outset, green architecture produces buildings that have long-term payoffs in using less energy and preserving more of the natural environment.

Brown pelicans at NOAA cleanup site after an oil spill near Tampa Bay. © NOAA Restoration Center, Louise Kane.

Zebra mussels invade the shores of the Great Lakes. © U.S. Environmental Protection Agency.

Recycling treated sewage (sludge). © U.S. Environmental Protection Agency.

Bleached boards from an old lumber mill operation are exposed amidst the wetlands vegetation at Grassy Point Wetland, an ecological restoration site at the mouth of the St. Louis River, Duluth, Minnesota, 1994. © U.S. Environmental Protection Agency.

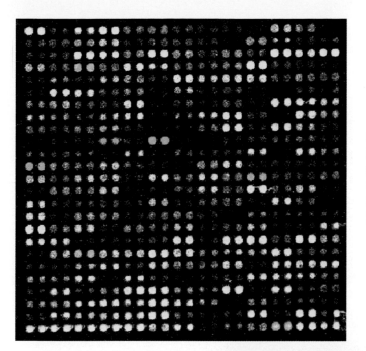

Gene array on "15K" chip (~15,000 spots) reveals susceptibilities to genetic disorders. © C. R. Schar & M. J. Doktycz / Oak Ridge National Laboratory / Department of Energy.

Paleontologist carefully chips rock matrix from a column of dinosaur vertebrae at the Dinosaur National Monument. These bones will be left in place for the visitor to see them just as they were deposited. © United States Geological Society.

Dust storm approaching Stratford, Texas, 1935. © NOAA, George E. Marsh Album.

Satellite image showing photosynthesis in oceans. © Orbital Imaging Corporation and processing by NASA Goddard Space Flight Center.

A clam shell bed around a thermal mound in the Pacific Ocean. © OAR / National Undersea Research Program (NURP); Univ. of Hawaii / NOAA.

Purple loosestrife (invasive species) growing in Saginaw Bay, Lake Huron, 1994. © U.S. Environmental Protection Agency.

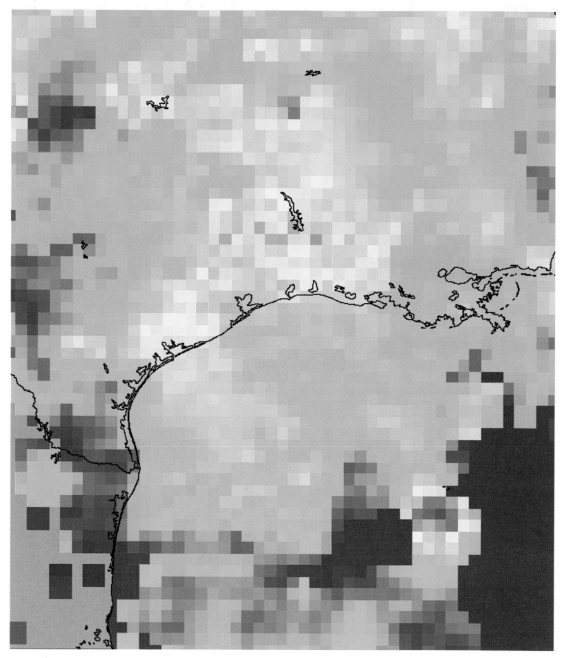

Satellite remote sensing provides a useful way to investigate the impact of intense local pollution sources, such as urban and industrial emissions, on regional air quality. This false-color image shows carbon monoxide plumes at an altitude of roughly 2 km (700 millibars) in the atmosphere over the Houston area and extending out over the Bay of Galveston. © NASA Newsroom/David Edwards.

Balloon launch at dawn near the South Pole to gather ozone data, 2002. © Sam Oltman / NOAA.

Balloon launch during the daytime near the South Pole to gather ozone data, 2002. © Sam Oltman / NOAA.

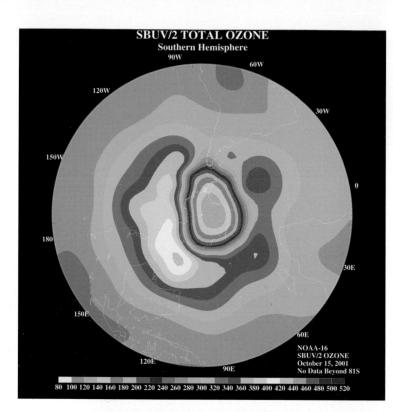

NOAA has used GIS technology to map changes in the ozone "hole" over the Antarctic. The ozone hole data is in units of ozone in each location, but on GIS maps it appears as purple, red, burgundy, and gray areas that have appeared over Antarctica since the fall of 2000. © Craig Long / NOAA.

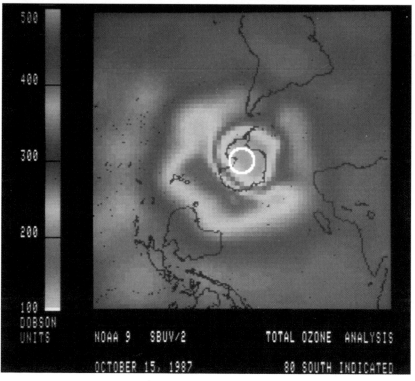

Ozone analysis map showing "ozone hole" over Antarctic continent, 1987. © NOAA.

For some large animals, even minimal human encroachment may be too much, as they need enormous ranges for maintaining breeding populations. Grizzly bears, for instance, need a huge home range (50–300 square miles for females; 200–500 square miles for males). They live in forests, moist meadows, and grasslands in or near mountains. A proposed wildlife corridor of such habitat between the Yukon in Canada and Wyoming's Yellowstone National Park may provide the needed territory without causing too much intrusion on areas occupied by human populations. Similar corridors are contemplated for other large animals.

In the United States, cars and trucks are an important part of life, but they use a great deal of energy and contribute to pollution. In the past little was done about this except in times of scarcity, such as the 1975 oil crisis. In record time, manufacturers began producing cars that obtained many more miles per gallon of gasoline. After the crisis passed, though, we returned to our former gas-guzzling ways, eventually buying sport-utility vehicles and similar vehicles with low mileage. Now there are some signs of hope: Battery-powered electric cars have been produced for short trips. These quiet cars emit no air pollution (though the electricity on which they run may have been produced from coal) and are popular for some purposes, but the need to recharge the battery limits their use. Hybrid gasoline-electric vehicles are also appearing (Toyota and Honda both make hybrid models); as they recharge themselves, they are more convenient and can be used for longer trips, but they do cause some pollution. Designers are also beginning to think about using hydrogen fuel cells to produce electricity for powering cars. The only product of combustion in a hydrogen-powered car would be water. General Motors has already brought out a prototype car that can be ready for mass production by 2010.

Cleaning up Pollution

Humans have added pollutants of all kinds—pesticides, fertilizers, industrial chemicals, medicines—to nearly every environment in the world. Some can be removed by simple, mechanical procedures; others require chemical treatment; still others seem destined to remain forever. One new approach to removing pollution, called bioremediation, makes use of biological treatment.

Benzene, a dangerous carcinogen, is used as a solvent in manufacturing processes, such as the production of paints. It is also found in vehicle exhaust fumes. Military bases that are now being converted to civilian uses often have a lot of benzene in the soil. Because benzene is slightly soluble in water and does not break down easily, once released into the environment it is hard to remove.

Some bacteria, such as *Pseudomonas,* feed on benzene and other hydrocarbons and can be used for cleaning up oil spills. As long as they have access to oxygen, that technique works, but in some cases the material is cut off from air. *Pseudomonas* and most other organisms that consume hydrocarbons are aerobic (they need oxygen to live).

What is needed are anaerobic bacteria that feed on hydrocarbons. John Coates and others at Southern Illinois University have now isolated some anaerobic bacteria (strains of *Dechloromonas*) that use benzene. Research is continuing on other bacteria that can be used for bioremediation. It is one of the most promising areas of waste treatment.

This ability of bacteria to adapt quickly to new environments is a two-edged sword:

It can be dangerous to humans who are overusing antibiotics, but it may prove highly useful in our struggle to clean up pollution.

The Earth Tomorrow

What will future life on Earth be like? For many years, pessimists have warned that greedy humans will destroy ourselves and much else on the planet; that we will create a long, dark nuclear winter in which only the bacteria and cockroaches are likely to survive.

Optimists, in contrast, have forecast a sunny future made possible by the wise use of technology, in which most people can live long, healthy lives in a favorable environment. Human population size will be voluntarily controlled, and biodiversity will be maintained by sustainable agriculture and ecosystem management.

Probably neither vision is entirely accurate. In the future as in the past, many humans are apt to make enormous mistakes regarding health and life, errors that will lead to the loss of human life and the continuing loss of other species. But we may muddle through again, given some science education and some thought about the consequences of our actions. As late astronomer Carl Sagan (1934–1996) once said, "It is suicidal to create a society dependent on science and technology in which hardly anybody knows anything about science and technology."[18] If we continue learning about the biosphere and how we interact with it, we may not only compensate for our mistakes but also build a world that can support ourselves and other living things for the centuries to come.

NOTES

1. Michael Balter, "Speech Gene Tied to Modern Humans," *Science* 297 (August 16, 2002): 1105.

2. Plague and Public Health in Renaissance Europe Project, www.iath.virginia.edu/osheim/plaguein.html.

3. "For the Ozone Layer, a New Look," *New York Times,* November 8, 2002.

4. J. M. Kiesecker et al. "Complex Causes of Amphibian Population Declines," *Nature* 410 (April 2001), 681–84.

5. Katherine Hunt, "Polluted Clouds May Give Patchy Cooling to World," Reuters News Service Online (Planet Ark), www.planetark.org/dailynewshome.cfm.

6. "Five Fast Breeder Reactors Planned," *Hindu,* March 7, 2003, www.hinduonnet.com/thehindu/2003/03/07/stories/2003030709000700.htm.

7. Sierra Club of Canada, www.sierraclub.ca/national/nuclear/reactors/iter-fusion-background.html.

8. Roderick Nash, *Wilderness and the American Mind* (New Haven: Yale University Press, 1973).

9. C. Wittwer, "The 1908 White House Governors' Conference," *Environmental Education* 1 (1979): 142.

10. Aldo Leopold Foundation, www.aldoleopold.org.

11. Marston Bates, "The Human Environment," speech given at University of California School of Forestry, April 23, 1962.

12. Tony McMichael, "The Biosphere, Health, and 'Sustainability,' " (Editorial) *Science* 297 (April 2002), 1105.

13. E. O. Wilson. Interview on "Talk of the Nation," National Public Radio, January 14, 2002.

14. Nigel C. A. Pitman and Peter M. Jorgensen. "Estimating the Size of the World's Threatened Flora," *Science* 298 (2002): 989.

15. J. Kirchner, telephone interview by author, Alamedo, Calif., 5 January 2003.

16. J. Robbins, "Outbreaks of a Rare Botulism Strain Stymie Scientists," *New York Times,* October 22, 2002.

17. U.S. Geological Survey Patuxent Wildlife Research Center, www.pwrc.usgs.gov/btreesnk.htm.

18. Quoted in T. J. Sejnowski, "Tap into Science 24–7," *Science* 301 (August 1, 2003): 601.

Investigations

You can explore many areas of the living world by carrying out the investigations that follow. Some are provided as concrete examples of ideas in the text; some will help you plan and carry out biology experiments; some will link the information in this book to events in your environment.

Before doing any investigations, be sure you have a spiral-bound notebook to be used only for your lab notes. Write your name on it. It is for your personal use only; if you are working with another person, you should each have a notebook. Write all your calculations and results in that book, being sure to date each page as you use it. Use it for any notes or questions to yourself, also. It may get messy with use (especially outdoors), but try to keep it readable!

Plan each investigation before beginning it. You won't need any elaborate equipment, but you may need to assemble some things or buy some inexpensive items. Have them all in your work area, which should be set aside for your use until you finish the investigation. A table in a basement or garage can be a good place to work.

When you finish an investigation, write up a formal report at least one page in length. Each report should have four major sections, with these headings:

Purpose. This section should be only one or two sentences long and should answer the question of why you are doing this investigation. The purpose might range from "I wanted to observe and identify the organisms in a drop of pond water" to "I wanted to test my hypothesis that calcium inhibits transmission of nerve impulses in frog neurons." If the lab instructs you to form a hypothesis, include it in this section.

Methods. Describe your procedures exactly, so that another person could do the same study by following your directions. List any equipment in this section.

Results. What did you find out or observe? Include any measurements and calculations here. (The original measurements and calculations should be in your notebook.) In some cases you may want to summarize your results in the form of a table. In addition to being convenient for summarizing, a table sometimes reveals patterns in the results that are not apparent from raw data.

Discussion. How do you interpret your results? For example, if you found that butterflies seemed to avoid a plant that you had hypothesized would attract them, why do you think that happened? Can you write a new, testable hypothesis about it?

Some investigations include questions for you to answer. Write your answers in this part of your report. (The answers are given at the end of this chapter.)

THE BIOSPHERE

The investigations in this section should be carried out in connection with Chapter 1. Some background information and terminology in that chapter may be helpful.

Population Stats

Find a lawn that has not been mowed recently. Throw a coin onto the lawn at random. Measure and record the height of the blade of grass nearest the coin. Repeat this procedure until you have recorded the heights of 30 blades of grass.

For recording your results in your report, group the data in five columns of a table like this:

	Height in cm				
	0–1	>1–2	>2–3	>3–4	>4–5
Number of blades	5	7	10	5	3

In this example, the shortest blades were 0–1 cm high, and the tallest were >4–5 cm high.

In your report, draw a bar graph representing the same data. State the range and average height. (If you need to, refer to Chapter 1.)

Examine Figure 7.1, which shows average blood sugar levels that might have

Figure 7.1
Blood Sugar Levels in a Hypothetical Population in 1980.

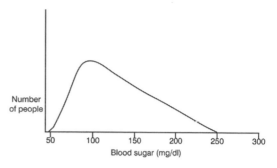

been found in a certain U.S. population of 100 persons in 1980. The x axis shows blood sugar level, and the y axis shows the number of people having that level.

1. What is the range of blood sugar levels in this population?

2. What is the average blood sugar level in this population?

3. Suppose that the same kind of data was collected for a similar population in 2000, and that a graph of the data looked like Figure 7.2.

How does Figure 7.1 differ from Figure 7.2? Suggest some factors that might explain the difference.

Figure 7.2
Blood Sugar Levels in a Similar Population in 2000.

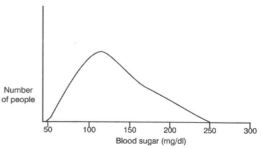

Soil Community

Construct a terrarium using soil and organisms from your own area. (An old aquarium, or any large glass container, is fine for this purpose.) To provide drainage, put a two-inch layer of gravel on the bottom. Add a few inches of soil, and plant some local plants in it. Place a few rocks on the soil. Finally, add some small animals—insects, spiders, snails, perhaps a very small salamander or lizard.

Place the terrarium in a location that receives light, but not bright sunlight. Water the soil sparingly; overwatering a terrarium is very easy. (Spraying the soil's surface lightly with a mister is better than pouring water onto it.)

Prepare an index card for each kind of plant and animal in the terrarium. Write the common name, genus, and species (use guides or the Internet to find out what these are) at the top of the card. By observing the organisms in the terrarium, add information about what they eat or are eaten by. If any organisms die, remove them after observing what happens to them.

Write a short essay as your report, describing the interacting organisms and your observations.

Life in an Aquarium

Set up a freshwater aquarium. Ideally, you will use an aquarium holding at least 10 gallons of water. (It is especially important to have a large water surface area so that enough oxygen can enter the water from the air.)

Place the empty aquarium where you plan to keep it, as you will not be able to move it after it is filled with water. It will need a little light, but not bright sunlight. Often an east window provides the best amount of light.

Setting up an aquarium should be a gradual process. You are creating an environment, and each element in it affects the others. So add one thing at a time, and wait a day or so before adding anything else. You will need to begin with a layer of aquarium gravel. Before adding it to the aquarium, rinse the gravel several times. When the rinse water is clean, you can put the gravel in the aquarium.

Add a few well-rinsed rocks, also. These are mainly for decoration; they should be no more than a few inches long or wide.

Plant a few aquarium plants in the gravel before adding any water. In your notebook, keep a record of the plants' names.

The water should be aerated before it is added to the aquarium, so that any chlorine (which can kill fish) will evaporate. Pour water into some large containers and leave it overnight before pouring it into the aquarium. You can add some floating aquarium plants now if you wish.

Add a few animals, such as snails and small fish. Record the common and scientific names of all the organisms in your notebook.

Feed the fish very lightly; it is easy to overfeed them. At each feeding, observe any events in the aquarium. Is the water changing color? Are the animals eating the plants or each other?

From your observations, diagram the food web in the aquarium. Diagram the oxygen–carbon dioxide cycle, also, in terms of the organisms you used. Write a paragraph in your report describing decomposition in this ecosystem.

THE EVOLUTION OF LIFE

The investigations in this section should be carried out in connection with Chapter 2. Some background information and terminology in that chapter may be helpful.

Jelly Bean Evolution

Materials: 4 oz. or more jelly beans (at least 15 different flavors)

Procedure: Using the jelly beans to represent fossil and modern organisms, make a model of evolution in a group of organisms. Assume that the jelly beans' exterior appearance accurately reflects their genes. Examine the jelly beans' colors and patterns. Notice whether the colors are solid, striped, or something else. Arrange them on a flat surface to form a phylogenetic tree. The "ancestral bean" at the base must be white, but otherwise the arrangement is up to you. Most of the beans along the branches should represent fossils; at the end of each branch is, today's group. The branches must follow some logical order, such as white to pink to red to speckled red. If a bean seems to have no relationship to the others, it can be placed by itself at the end of a long branch, with a hypothetical missing link between the white ancestor and the other gene.

A Phylogenetic Tree

Think about what you know about the following organisms: *E. coli* bacteria, dogs, wolves, dogwood trees, and tigers. Sketch a phylogenetic tree representing their evolution. Which are most closely related? These should be closest in the diagram. Which are only distantly related? These should be farther apart. In your report, draw the tree and write a paragraph explaining why you drew it as you did.

Upsetting Hardy and Weinberg

Use 200 real red and white beans to represent genes in a population of 100 organisms. Twenty-five percent of the population (p^2) is homozygous red *(RR)*, represented by two red beans. Fifty percent of the population (2 *pq*) is heterozygous red

(Rr), represented by one red and one white bean. Twenty-five percent of the population (q^2) is homozygous white *(rr),* represented by two white beans. (The hypothetical white trait is inherited as a recessive trait.) To represent the three groups, place 25 pairs of red beans together, 50 red and white pairs together, and 25 pairs of white beans together.

Assume a colorblind predator eats all of the paired white beans, but he can't see the color red. Remove the paired white beans. You now have 150 beans in the gene pool. What proportions of homozygous red, heterozygous red, and homozygous white remain? Now assume that the remaining organisms mate at random and produce a new generation. Again, the predator eats all the paired white beans, and the remaining organisms mate at random and produce a new generation. How many generations of random mating and predation do you think it will take to remove the white allele from the population gene pool?

CELLS AND GENETICS

The investigations in this section should be carried out in connection with Chapter 3. Some background information and terminology in that chapter may be helpful.

Modeling Mitotic Division

Use the description of mitosis in Chapter 2 to create "chromosomes" made of beads, pipe cleaners, or other materials. (What kind of materials will be most useful?) Then use the chromosomes to demonstrate the process to another person.

You Be the Judge

If paternity must be established today, DNA fingerprinting makes it quite easy. Not many years ago, though, a verdict that a man had fathered a certain child had to be deduced from other genetic information.

Suppose a woman sued a man for child support, insisting that he was the father of her brown-eyed child. As the judge, you notice both the man and woman have pale blue eyes. How would you rule? What is your reason?

Another woman sued a man who was colorblind, saying that he must be the father of her colorblind son. The woman herself had normal vision. How would you rule? Why did you make that decision?

Assembling DNA

Draw geometric shapes representing the four nucleosides in DNA, making sure that the A and T bases fit together and the C and G bases fit together. Make two long phosphate backbones with paper tape. Make many copies of the nucleosides. (These can be copied on colored paper, or you can color them.) Use these to assemble a mobile, being sure to have the two strands arranged in opposite directions ($5'$–$3'$ and $3'$–$5'$). Finally, coil it around a straightened coat hanger or dowel rod, and hang the helical model in your study area.

You may think this is a rather childish investigation, but it is not too different from the method Watson and Crick used; also, making this model successful ensures that you understand DNA's stereochemistry.

A Royal Disease

Several of Europe's royal families carry the sex-linked gene for hemophilia. Hemophiliacs' blood does not clot normally, and they are in danger of bleeding too much from even small cuts. Carriers of the gene are not affected.

Assume that in one of these royal families, a princess is normal with respect to hemophilia, but one of her brothers has the disease. A prince in another country who wants to marry her is concerned about the potential for hemophilia in their children. No one in his family has had the disease for several generations, and you can assume he is free of the gene. Construct a likely pedigree for the prince and princess and the five children (three boys and two girls) they may have.

Genealogy

Assemble all the information available about your own pedigree by interviewing relatives. Find out what you can about any inherited diseases, but include other diseases also (in case they turn out someday to have a genetic component). Construct your personal pedigree and annotate it with any comments about diseased persons that might be useful in making medical decisions for yourself or your children.

In many families this may be a sensitive topic; if you suspect a relative is hiding information about parentage or diseases, do not press them. At the same time, try to emphasize the importance for everyone of knowing about illnesses that may have a genetic component. If you have not alienated them at the outset, relatives who are reluctant to give you information may decide to do so after they have time to think it over. If one relative cannot or will not help you, you may be able to fill in the blanks by talking with other relatives.

ORGANS AND SYSTEMS

The investigations in this section should be carried out in connection with Chapter 4. Some background information and terminology in that chapter may be helpful.

Using a Microscope

Borrow a compound microscope and some prepared slides of plant and animal tissues if possible. Even an inexpensive instrument can show quite a bit of detail. Using the figures in Chapter 3 as guides, try to identify the nucleus, cytoplasm, chloroplast, cell membrane, and cell wall if these structures are present.

A compound microscope makes it possible to magnify the image of an object twice—once with the lens in the eyepiece (at the top of the microscope), and again with the low- or high-power lens just above the object of study. Many low-power lenses magnify objects $10\times$, and many high-power lenses magnify them $21\times$. To find out the total magnification, multiply the number on the eyepiece (often $10\times$) by the number on the lens you are using. For instance, using a $10\times$ eyepiece with a $21\times$ lens would give a total magnification of $210\times$. In your results section, state the total magnification for each picture.

Look at some living cells, also. (Use the kind of glass slide that has a depression or well in the center, so that you can add water.) For examining these cells, keep them suspended in water to keep them from drying out. Sources of interesting cells are all around you: You can find plant cells in ponds or aquaria. Live-culture yogurt contains yeast cells. You can carefully scrape epithelial cells from the inside of your cheek, using a very clean teaspoon.

Use your imagination to decide what other kinds of cells you would like to examine. You may enjoy looking at mushrooms, onion skin, or insects. Whatever you look at, draw a picture of it in your notebook and label the parts you can identify. For your lab report, copy the drawings from your notebook or redraw them.

For comparison, obtain some pictures of cells that were made with electron microscopes. (*Scientific American* magazine is a good source of electron micrographs.) At what magnification were the pictures made? What cell structures can be seen with the electron microscope that can't be viewed with a light microscope? What advantages might each kind of microscope have?

Integrated Systems

Caution: This activity will make the heart beat faster and will cause some stress on the skeleton. *If you have any problem that may be worsened by the activity, do not carry it out yourself.* You may want to have another person run or jump as directed while you record the measurements and write up the report.

Measure your pulse rate by placing the index and middle fingers of one hand on the pulse at your wrist or beneath your chin. Count the number of beats for 15 seconds, as shown on a stopwatch or other timer. Multiply the number by four to determine your initial pulse rate. Record the number.

Now, run in place or skip rope for two minutes. As you do so, pay attention to your breathing and how your leg muscles and ankles feel. You may be conscious of your heartbeat, and you may start to sweat.

When you stop running, immediately measure and record your pulse rate again. In your notebook, write some notes about the physical responses you felt.

Write an essay or construct a labeled diagram to explain how running or skipping rope affects the body's various systems. Include comments about the need for energy and how it is obtained, and use as much chemistry as you can in your explanation.

Nutrition

Following a balanced vegetarian diet may benefit both your own health and the environment: Vegetarians are less likely than meat-eaters to be obese or to have diabetes or cardiovascular disease; and by moving down one step in the food chain vegetarians can be supported by fewer organisms. Because the growing human population is consuming more of the world food supply while removing some ranges and agricultural land for housing developments, providing enough food for everyone will eventually be difficult. For that reason, many people who are concerned about overpopulation and the environment are vegetarians. Many vegetarians are concerned about the effects of meat on their health, especially if the meat contains antibiotics or hormones. Some vegetarians object to killing other animals for food.

Plan a nutritious diet for a vegetarian (specify the type) for one week. It must include all the essential amino acids, oils, vitamins and minerals provided by a normal diet that includes meat and dairy foods.

Optional: Use the diet yourself for a week. Write an essay about how you felt while on the diet and how this diet relates to environmental issues.

Plant Physiology

Find out about the effects of alternating darkness and light on plants. You will need:

- six mature, healthy aster or chrysanthemum plants of the same size that have not quite reached the flowering stage.
- two plant-growing lamps with timers.

Place three of the plants under one of the lamps and three under the other. Each set should be put in separate closets or other spaces where the lamps will be the only source of light. Water them daily with just enough water to keep the soil surface moist. Set one of the lamps on a cycle of 8 hours of light and 16 hours of darkness daily. Set the other on a cycle of 16 hours of light and 8 hours of darkness daily. Make a hypothesis predicting what you think will happen to the two sets of plants. Observe the plants

daily for three to four weeks, taking notes on their appearance.

In your report, include comments on whether the asters or chrysanthemums are short-day or long-day plants, and on what advantages this adaptation (called photoperiodism) may have for the plants. How do florists take advantage of it?

Circulatory System

Make a photocopy of a good drawing of the human circulatory system. On your copy, use blue and red pencils to color the areas of unoxygenated (blue) and oxygenated (red) blood. In your report, write a paragraph about hemoglobin's role in carrying oxygen and how hemoglobin itself is changed chemically in that process.

The Heart of the Matter

Obtain a steer's heart from a butcher. If you have no dissection instruments, use a single-edged razor blade to dissect the heart, but be very careful and have someone with you to help you. Slice carefully through the top layer of muscle from anterior to posterior and from left to right. Then fold back the flaps and look inside. Observe the parts, then draw and label them. Continue cutting and folding until you see every part of the heart shown in Figure 4.2. On the drawing, diagram the flow of blood through the heart, using blue arrows to show the direction in which unoxygenated blood flows and red arrows to show the direction in which oxygenated blood flows.

A SURVEY OF ORGANISMS

The investigations in this section should be carried out in connection with Chapter 5. Some background information and terminology in that chapter may be helpful.

Plant Keys

Obtain a guide to identification of a group of local plants that is based on a dichotomous key (a scheme in which you begin with a general description of a plant to be identified and proceed through a series of either/or choices until you reach the plant's name). These small guides, often available in libraries and sold in museums and recreation-area stores, are sometimes called finders. Some examples are Tom Watts's *Rocky Mountain Tree Finder* and May Thielgard Watts's *Flower Finder.*

Use the finder to key or identify about 25 local plants. If you have trouble deciding which choices to make, look up the problem area in a botany textbook. Sometimes you can try choosing one path and seeing where it leads you. If you end up at the name of a plant you have already identified by other means, you can retrace your steps through the key and find out more information about your plant.

Plants in Your Grocery Store

Assemble at least 10 different fruits and vegetables and some mushrooms that are sold as foods. Identify as many as possible of the following plant parts on them: leaves, roots, stems, spores, seeds, flower petals, seed pods.

Backyard Classification

Make a survey of plants and animals in your neighborhood. Set up a classification system or a dichotomous key for them. Ignore standard schemes, just use the organisms you have. What characteristics did you use?

HUMANS IN THE BIOSPHERE

The investigations in this section should be carried out in connection with Chapter 6. Some background information and terminology in that chapter may be helpful.

Environmental Controversies

Browse the Web and other media to obtain information on a current environmental controversy that interests you, such as the use of Colorado River water by several states. Construct a table to summarize your findings about individuals or groups involved, the actions they recommend, their reasons for the recommendations, their evidence, and your additional comments (such as the reliability of the individual or group). Write a paragraph giving your own recommendation about the issue, based on what you have learned.

Museum Exploration

Visit a nearby natural history museum or find an online virtual museum where you can see a display about your local environment. Find out about the geology—the rocks, soil, and climate—as well as the biology. In your report, describe the biome(s) where you live in terms of its geology and biology. Also, list the ways in which humans have changed the natural ecosystem where you live.

Population Growth

Examine the table below, which shows changes over several decades in U.S. state and national populations. Has your state's population grown or declined since 1980? Construct a graph that compares the population changes in your state since 1900 with changes in the national population. In your report, discuss how national and state population changes have affected aspects of daily life and what may have led to the population changes.

Planting a Butterfly Garden

Most people are drawn to butterflies by their beauty and grace, but biologists find them intriguing for other reasons. Nineteenth-

United States Population Figures

State	Population in 1900	Population in 1950	Population in 1980	Population in 2000
Alabama	1,829,000	3,062,000	3,894,000	4,447,100
Alaska	64,000	129,000	402,000	626,932
Arizona	123,000	750,000	2,718,000	5,130,632
Arkansas	1,312,000	1,910,000	2,286,000	2,673,400
California	1,485,000	10,586,000	23,668,000	33,871,648
Colorado	540,000	1,325,000	2,890,000	4,301,261
Connecticut	908,000	2,007,000	3,108,000	3,405,565
Delaware	185,000	318,000	594,000	783,600
District of Columbia	279,000	802,000	638,000	572,059
Florida	529,000	2,771,000	9,746,000	15,982,378
Georgia	2,216,000	3,445,000	5,463,000	8,186,453
Hawaii	154,000	500,000	965,000	1,211,537
Idaho	162,000	589,000	944,000	1,293,953
Illinois	4,822,000	8,712,000	11,427,000	12,419,293
Indiana	2,516,000	3,934,000	5,490,000	6,080,485
Iowa	2,232,000	2,621,000	2,914,000	2,926,324
Kansas	1,470,000	1,905,000	2,364,000	2,688,418
Kentucky	2,147,000	2,945,000	3,661,000	4,041,769
Louisiana	1,382,000	2,684,000	4,206,000	4,468,976
Maine	694,000	914,000	1,125,000	1,274,923
Maryland	1,188,000	2,343,000	4,217,000	5,296,486
Massachusetts	2,805,000	4,691,000	5,737,000	6,349,097
Michigan	2,421,000	6,372,000	9,262,000	9,938,444
Minnesota	1,751,000	2,982,000	4,076,000	4,919,479
Mississippi	1,551,000	2,179,000	2,521,000	2,844,658
Missouri	3,107,000	3,955,000	4,917,000	5,595,211
Montana	243,000	591,000	787,000	902,195
Nebraska	1,066,000	1,326,000	1,570,000	1,711,263
Nevada	42,000	160,000	800,000	1,998,257
New Hampshire	412,000	533,000	921,000	1,235,786
New Jersey	1,884,000	4,835,000	7,365,000	8,414,350
New Mexico	195,000	681,000	1,303,000	1,819,046
New York	7,269,000	14,830,000	17,558,000	18,976,457
North Carolina	1,894,000	4,062,000	5,882,000	8,049,313
North Dakota	319,000	620,000	653,000	642,200
Ohio	4,158,000	7,947,000	10,798,000	11,353,140
Oklahoma	790,000	2,233,000	3,025,000	3,450,654
Oregon	414,000	1,521,000	2,633,000	3,421,399
Pennsylvania	6,302,000	10,498,000	11,864,000	12,281,054
Rhode Island	429,000	792,000	947,000	1,048,319

(continued)

United States Population Figures *(continued)*

State	Population in 1900	Population in 1950	Population in 1980	Population in 2000
South Carolina	1,340,000	2,117,000	3,122,000	4,012,012
South Dakota	402,000	653,000	691,000	754,844
Tennessee	2,021,000	3,292,000	4,591,000	5,689,283
Texas	3,049,000	7,711,000	14,229,000	20,851,820
Utah	277,000	689,000	1,461,000	2,233,169
Vermont	344,000	378,000	511,000	608,827
Virginia	1,854,000	3,319,000	5,347,000	7,078,515
Washington	518,000	2,379,000	4,132,000	5,894,121
West Virginia	959,000	2,006,000	1,950,000	1,808,344
Wisconsin	2,069,000	3,435,000	4,706,000	5,363,675
Wyoming	93,000	291,000	470,000	493,782
Total United States	76,212,000	151,326,000	226,546,000	281,421,906

century entomologist Henry Walter Bates, writing of his work along the Amazon River, said that all of nature is revealed in the colored patterns of butterfly wings, which

vary in accordance with the slightest change in the conditions to which the species are exposed. It may be said, therefore, that on these expanded membranes Nature writes, as on a tablet, the story of the modifications of species, so truly do all changes of the organization register themselves thereon. Moreover, the same color-patterns of the wings generally show, with great regularity, the degrees of blood-relationship of the species. As the laws of Nature must be the same for all beings, the conclusions furnished by this group of insects must be applicable to the whole organic world; therefore, the study of butterflies—creatures selected as the types of airiness and frivolity—instead of being despised, will some day be valued as one of the most important branches of biological science.[1]

Indeed, biologists today still do find butterflies valuable subjects for study.

Whether you want to study butterflies or merely enjoy them, you may want to plant a butterfly garden. As shopping malls and housing developments expand and replace the fields and woodlands that are natural habitats for these insects, many of the species are dying out. Providing appropriate foods for butterflies in your garden can help perpetuate them.

Adult butterflies feed on nectar of flowers on a great variety of plants—goldenrod, thistle, asters, black-eyed Susans, white clover, Michaelmas daisies, mustards, yarrow, and sweet William, for example. The flowers that attract them tend to be orange, pink, yellow, purple, or white. Having some of those flowers around, especially in large groups, will probably bring butterflies into your garden. If you choose flowers that bloom at different times of the year, you will give the adults a continual supply of nectar. Some butterflies prefer feeding at greater heights than others do, so having flowers at various heights is helpful. Full sunlight for 8–10 hours a day, and a water supply (or even mud) are also needed.

For helping these insects produce the next generation, though, you need to provide food for the larvae, or caterpillars, as well; each species tends to prefer a certain plant for larval food. The female butterflies deposit eggs on that kind of plant, and the eggs grow into larvae, which then feed on the plant. When the larvae have eaten enough and are ready to develop further, they become pupae, the next stage in their life cycle. Pupae are somewhat protected by a chrysalis. A pupa may emerge as an adult in a short time, or may spend the winter in that stage.

The table below shows foods that are needed by the larvae of just a few common butterfly species. It is important to provide both plants that provide nectar for the adults and other plants that are used as food by larvae. A nursery can help you choose plants that are native to your area and that will attract your native butterflies.

Foods for Butterflies

Species	Common Name	Plants Used as Food by Larvae
Danaus plexippus	Monarch	*Asclepias tuberosa* (butterfly weed), *A. syriaca* (common milkweed)
Euphydryas sp.	Checkerspot	*Penstemon barbatus* (beard tongue), *Plantago* (plantain)
Vanessa virginiensis	Painted lady	*Echinops ritro* (globe thistle), *Malva alcea* (hollyhock mallow)
	Blues	*Lupinus* sp. (lupine), clover, dogwood, vetch, sorrel
	Fritillaries	*Viola* sp. (violets)
Precis coenia	Buckeye	*Plantago* (plantain)
Papilio cresphontes	Giant swallowtail	*Citrus sinensis* (orange)

ANSWERS TO QUESTIONS

Population Stats

1. 70–250 mg/dl.
2. 100 mg/dl.
3. In 2000 the U.S. population as a whole was more overweight than in 1980, and people got less exercise and ate more fattening foods. This combination can contribute to diabetes and other conditions in which there is an elevated blood sugar level. You may suggest other contributing factors.

Jelly Bean Evolution

There is no right answer, but there may be many wrong answers. There should be no abrupt changes from one color to another.

A Phylogenetic Tree

Your tree might look like this:

Figure 7.3
Phylogenetic Tree.

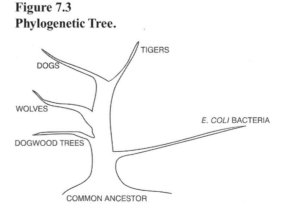

It is based on the ideas that *E. coli* probably evolved earlier than the other organisms, and is closest to their common ancestor; that dogs and wolves are very closely related; and that all animals came from a common ancestor. Genetic analysis of these organisms could yield a much more accurate diagram.

Upsetting Hardy and Weinberg

After the first predation, $p = 100/150 = 0.67$; $q = 50/150 = 0.33$.

$$p^2 + 2\,pq + q^2 = (0.67)^2 + 2(0.67)(0.33) + (0.33)^2$$

The frequency of white offspring (p^2) is now 0.11, or 11 out of 100. The frequency will continue to drop with continued predation, but the white allele will never be completely eliminated.

As this investigation shows, it is extremely difficult to remove a recessive gene from a population gene pool. Because it is masked by the dominant gene, in many cases there is no selection against it.

Modeling Mitotic Division

The best materials for this purpose are pop beads in two colors, because they can be used for chromosomes of maternal and paternal origin and of different lengths. These beads are not always available, and you may have to use more imagination in compensation.

You Be the Judge

It is nearly impossible for a man and woman with pale blue eyes to have a brown-eyed child. They both have the genotype *bb,* as will their children, and the *B* gene is needed for brown eyes.

The recessive gene for colorblindness is carried on the X chromosome. The son would have inherited his X chromosome from his mother, who must have carried the gene.

A Royal Disease

The pedigree should show the princess and prince as free of the disease. However, the princess is a carrier, as shown by one of her brothers being afflicted with hemophilia. Her daughters and sons will each have a 50% chance of inheriting the gene. Because the daughters will inherit the normal gene from their father, they will be normal (but may be carriers). The sons will have a 50% chance of having the disease.

Integrated Systems

During running and other aerobic exercises, your body calls on all of its systems to provide the muscles with energy. Your essay or diagram should include detailed information on the following:

- The respiratory system must speed up breathing to take in oxygen and lose carbon dioxide more rapidly.
- The circulatory system carries oxygen and glucose to the muscle cells and carries carbon dioxide away. The heart beats faster, speeding up circulation.
- The neuroendocrine system supplies adrenaline to other systems.
- The skin's sweat glands excrete sweat onto the body surface, cooling the body.
- With repeated exercise, the skeletal system is strengthened.
- The muscle cells' myosin and actin molecules move, and the mitochondria break down sugars for energy.
- The digestive system slows down temporarily as the result of adrenaline being secreted into the blood.

Nutrition

You can safely follow either a lacto-ovo-vegetarian (without meat) or vegan (without meat, dairy foods, or eggs) diet, but you need to plan so that you can get the nutrients ordinarily provided by a complete diet. Your daily menus should include:

- 0.45 g protein (450 mg), which can come from such sources as soy foods and beans. Recent research shows that it's not even necessary for complementary amino acids to be consumed at the same meal, as long as they are consumed on the same day. Eating plant proteins from varied sources seems to provide what is needed.

- 5,000 IU vitamin A, in sweet potatoes, cantaloupe, carrots, spinach, broccoli, winter squash.

- 60 mg vitamin C, found in citrus fruits, cabbage, broccoli, green pepper. In addition to being an important vitamin, vitamin C may help in the absorption of iron.

- 1.5 mg thiamin, from soy foods, enriched wheat flakes, dried white beans, brown rice.

- 1.7 mg riboflavin, found in broccoli, spinach, wheat germ.

- 20 mg niacin, from peanuts, navy beans.

- 1.0 g calcium, in fortified orange juice and other supplements, turnip or collard greens, sesame seeds, soy foods.

- 3.0 μg vitamin D, often added to soy milk and some breakfast cereals. Vegetarians should expose their hands, arms, and face to the sun for about 10 minutes each day to synthesize vitamin D. If that isn't possible, diet can provide it. Only strict vegans need to be concerned about adding vitamin D to their diets, because the minuscule amounts needed are supplied to lacto-ovo-vegetarians by dairy foods.

- 18 mg iron, contributed in small amounts by legumes, dried fruits, bran flakes, sea vegetables, leafy green vegetables. Surprisingly, vegetarian diets actually provide more iron than nonvegetarian diets, but the iron in plant foods is poorly absorbed. In spite of that, the rates of iron-deficiency anemia are no greater in vegetarians than in other people.

- A small amount of linoleic and linolenic acids, available in corn oil, canola oil, olive oil, walnuts. Depending on total calorie intake and the type of oil consumed, about two to three teaspoons of oil daily supply enough linoleic and linolenic acids.

Much helpful, specific advice from the American Dietetic Association is available online at www.eatright.org/adap1197.html.

Backyard Classification

You might group animals by number of legs, presence or absence of fur, presence or absence of antennae, and so on. Plants might be classified by leaf shape, color of flowers, or other characteristics.

NOTE

1. Henry Walter Bates, *The Naturalist on the River Amazons* (1864; Berkeley: University of California Press, 1962).

Appendix 1

Investigating Biology on the Internet

Although thousands of biological Web sites can be found on the Internet, many of them are unreliable or are unlikely to remain available. The sites listed below all appear to be from reliable sources that will be accessible for many years.

The sites are classified in accordance with the chapters in this book. They are free unless noted, though some may require registration. Readers are encouraged to visit these sites on the Web, especially museums.

GENERAL

www.scitechdaily.com	SciTechDaily Review. Short news notes and links to journal articles on science and technology topics, revised frequently.
www.sciencenews.org	The online version of *Science News* magazine.
www.sciam.com	The online version of *Scientific American* magazine.
www.scienceonline.org	The online version of *Science* magazine (American Association for the Advancement of Science).
www.jama.com	The online version of *JAMA* (*Journal of the American Medical Association*).
www.magazine.audubon.org	The online version of *Audubon* magazine.

CHAPTER 1. THE BIOSPHERE

http://oceanexplorer.noaa.gov	Ocean exploration, from the National Oceanic and Atmospheric Administration (NOAA).
www.mip.berkeley.edu/mvz	The Museum of Vertebrate Zoology at the University of California at Berkeley.

www.wwfus.org/ecoregions	World Wildlife Fund map and detailed information about biomes of the world.
www.biology.arizona.edu/biochemistry/ biochemistry.html	An introduction to biochemistry; a University of Arizona site.

CHAPTER 2. THE EVOLUTION OF LIFE

www.amnh.org	The American Museum of Natural History in New York.
www.ucmp.berkeley.edu	The elaborate and beautiful Web site of the Museum of Paleontology at the University of California at Berkeley.
www.nps.gov./dino/dinos.htm	Dinosaur National Monument in Utah.
www.dinodata.net	Dinosaur information, shared by paleontologists around the world.
www.cmnh.org	The Field Museum of Natural History in Chicago.
www.pbs.org/evolution	The online version of the *Evolution* public television series.
www.ucmp.berkeley.edu/help/timeform. html	Geologic "time machine" from the Museum of Paleontology at the University of California at Berkeley.
http://exobiology.nasa.gov	NASA's research on exobiology.

CHAPTER 3. CELLS AND GENETICS

http://newton.dep.anl.gov/askasci/ mole00087.htm	Ask a Scientist Molecular Biology Archive, Department of Energy.
www.ncbi.nlm.nih.gov/disease/dmd.html	National Center for Biotechnology Information (National Institutes of Health).
www.lpch.org/diseasehealthinfo/ healthlibrary/genetics/trans.html	Medical Genetics. Lucile Packard Children's Hospital, Stanford University Medical Center.
www.biology.arizona.edu/human_bio/ problem_sets/blood_types/intro.html	A tutorial about blood types from the University of Arizona Biology Department.
http://gslc.genetics.utah.edu	Genetic Science Learning Center from the University of Utah.
www.genome.gov	The Human Genome Project, from the National Human Genome Research Institute.
http://ghr.nlm.nih.gov	Genetics Home Reference. Information about genetic conditions from the National Library of Medicine.
www.jewishgenetics.org	Center for Jewish Genetic Disorders.

CHAPTER 4. ORGANS AND SYSTEMS

www.ninds.nih.gov/health_and_medical/disorders	National Institute of Neurological Diseases and Stroke, part of the National Institutes of Health.
www.loni.ucla.edu	A digital atlas of the human brain, with additional information about functions. A University of California at Los Angeles site.
http://vesalius.northwestern.edu	Vesalius's atlas, from Northwestern University.
www.ama-assn.org/ama/pub/category/7140.html	The AMA atlas of the body.
www.bartleby.com/107	The Bartleby.com edition of Gray's *Anatomy*.
www.nlm.nih.gov/research/visible/visible_human.html	The Visible Human project at the National Institutes of Health.
http://wellnessletter.com/html/wl/wlfeatured.html	The UC–Berkeley Wellness Letter.
www.eatright.org/adap1197.html	American Dietetic Association. Vegetarian diets—position of the ADA.
www.healthfinder.gov	Reliable health information from the U.S. Department of Health and Human Services.

CHAPTER 5. A SURVEY OF ORGANISMS

www.sandiegozoo.com/virtualzoo/polarcam.html	San Diego Zoo, featuring Web cams.
www.npwrc.usgs.gov/resource/othrdata/chekbird/chekbird.htm	Bird checklists for regions of the United States, from the U.S. Geologic Survey (USGS).
http://biodidac.bio.uottawa.ca/index.htm	Thousands of photos and drawings of animals of the world, from the University of Ottawa.
www.nps.gov/seki/fire/fire_res.htm	A National Park Service site. More than 50,000 digital images of plants, animals, people, and landscapes.
www.botany.net/idb	Internet directory for botany sites, compiled by botanists in several countries.
www.primate.wisc.edu	Primate Information Net. Extensive information about hundreds of primate species from the University of Wisconsin.
http://animaldiversity.unmz.umich.edu	The Museum of Zoology at the University of Michigan. Extensive information about animals.
http://scitec.uwichill.edu.bb/bcs/bl14apl/bl14apl.htm	A college-level introduction to botany, offered by the University of the West Indies.

CHAPTER 6. HUMANS IN THE BIOSPHERE

www.prb.org	Population Reference Bureau statistics.
www.who.int/en	World Health Organization.
www.epa.gov	Environmental Protection Agency.
www.ncdc.noaa.gov	National Oceanic and Atmospheric Administration.
www.eia.doe.gov/neic	Department of Energy information center.
http://species.fws.gov	Fish and Wildlife Service.
www.usgs.gov	U.S. Geological Survey
www.lungusa.org/air2001/index.html	Air quality information from the American Lung Association.
www.epa.gov/iaq	EPA information about indoor air quality.
www.fao.org/forestry/fo/country/nav_world.jsp	Profiles of the world's forests from the Food and Agriculture Organization.
www.nps.gov/seki/fire/fire_res.htm	Fire ecology; a National Park Service site.
www.epa.gov/enviro/wme	The EPA's Window to My Environment.
www.epa.gov/epahome/commsearch.htm	EPA information about your local community.
www.nasm.si.edu/earthtoday	Digital views of Earth, from the National Air and Space Museum.
http://minerals.usgs.gov/minerals	Information about minerals.
www.audubon.org	National Audubon Society.
www.epa.gov/globalwarming/impacts	EPA site about the possible impacts of global warming.
www.frogweb.gov	Amphibian declines and deformities; a USGS site.

Appendix 2

Units of Measurement: Metric System and English System

In the United States we use the English system of measurement units for everyday purposes. In science, and in most of the world outside the United States, the Systèm International d'Unités (SI) is used for measurement. It is also called the metric system. The basic SI units are the meter (length), the kilogram (mass), the kelvin (temperature), and the second (time). They are modified by multiplying or dividing in multiples of 10, as follows:

LENGTH

1 kilometer (km)	=	1,000 meters
1 hectometer (hm)	=	100 meters
1 dekameter (dkm)	=	10 meters
1 decimeter (dm)	=	0.1 meter
1 centimeter (cm)	=	0.01 meter
1 millimeter (cm)	=	0.001 meter
1 micrometer (μm)	=	0.000001 meter
1 nanometer (nm)	=	0.000000001 meter

MASS

1 kilogram (km)	=	1,000 grams
1 hectogram (hm)	=	100 grams
1 dekagram (dkm)	=	10 grams
1 decigram (dm)	=	0.1 gram
1 centigram (cm)	=	0.01 gram
1 milligram (cm)	=	0.001 gram
1 microgram (μm)	=	0.000001 gram
1 nanogram (nm)	=	0.000000001 gram
1 microgram (μm)	=	0.000001 gram

VOLUME

1 kiloliter (km)	=	1,000 liters
1 hectoliter (hm)	=	100 liters
1 dekaliter (dkm)	=	10 liters
1 deciliter (dm)	=	0.1 liter
1 centiliter (cm)	=	0.01 liter
1 milliliter (cm)	=	0.001 liter
1 microliter (μm)	=	0.000001 liter
1 nanoliter (nm)	=	0.000000001 liter
1 microliter (μm)	=	0.000001 liter

The following table will enable you to convert SI units to English units and vice versa.

1 kg	=	2.2046 lb
1 g	=	0.035 oz
1 lb	=	0.454 kg
1 inch	=	2.54 cm
1 km	=	0.62 mi
1 meter	=	39.37 in
1 cm	=	0.39 in
1 ft	=	30.48 cm
1 foot	=	0.3048 meters
1 mile	=	1.6093 km

1 cubic centi-
meter (cm³)   = 1 ml (approximately)
32 degrees F   = 0 degrees C or 273 K

Any easy way of converting temperatures between Fahrenheit and Celsius is this three-step system:

1. Add 40 to the beginning temperature.

2. If you are converting from Fahrenheit to Celsius, divide by 1.8. If you are converting from Celsius to Fahrenheit, multiply by 1.8.

3. Subtract 40 to get the new temperature.

Appendix 3

Landmarks in Biology

ANCIENT BIOLOGY

Sixth century B.C.   Thales and other Ionian rationalist philosophers rejected supernatural explanations for natural phenomena.

Anaximander believed that life originated in the sea and that all land animals are descended from sea animals.

Heraclitus of Ephesus thought the original substance was fire and that everything begins and ends in fire.

Anaximenes of Miletust thought that air was the origin of all things.

Fifth century B.C.   Empedocles was one of the first to suggest that the entire universe was made up of four elements (earth, water, fire, and air) and was the first to put forward the idea that species had developed by evolution.

Leucippus and his pupil Democritus of Abdera were the first to hypothesize that all matter is composed of atoms.

Hippocrates introduced a rational approach to medicine and promulgated the concept of letting "natural law" effect cures.

Fourth century B.C.   Aristotle, the founder of zoology, described about 500 kinds of animals and classified them in a ladderlike *Scala Naturae;* emphasized direct observation and experimentation.

Threophrastus founded botany and described about 500 kinds of plants.

Second century B.C.   Galen described human anatomy, based largely on studies of other animals.

BIOLOGY IN MEDIEVAL TIMES

First century A.D.   Retreat from rationalism into preserving the teachings of the past.

Thirteenth century   Rediscovery of Aristotle by Albertus Magnus.

Thomas Aquinas's attempt to harmonize Aristotelian philosophy and Christian teachings.

Fifteenth century	Leonardo da Vinci dissected animals and humans to draw them accurately.
	Ships reached the New World and made studies of new plant and animal species possible.

BEGINNINGS OF MODERN BIOLOGY

1543	Vesalius wrote his textbook of human anatomy, the *Fabrica.*

Seventeenth Century

1628	William Harvey published his findings about the circulation of blood.
1640	Jan Baptista van Helmont showed plants did not get their sustenance from the soil and made the first studies of biochemistry.
1665	Robert Hooke published *Micrographia,* containing his descriptions and illustrations of cells, which he named.
1668	Francesco Redi disproved idea of spontaneous generation.
1670s	Anton van Leeuwenhoek built the first simple microscope.
	John Ray accumulated detailed descriptions of plants and animals and attempted to classify them.

EIGHTEENTH CENTURY

1735	Carl Linnaeus (the father of taxonomy) published *System Naturae* and devised a system for naming, ranking, and classifying organisms that is still used.
	Gilbert White described variations in plants and animals.
	Count Georges-Louis de Buffon described the struggle for existence, variations within species, and other matters that formed important parts of Darwin's theory.
1759	Caspar Friedrich Wolff described differentiation in plants.
1774	Joseph Priestley discovered oxygen.
1785	James Hutton, the founder of historical geology, first proposed that the Earth's crust alternately was uplifted and subsided because of dynamic forces acting within it during geologic time periods.
1791	Luigi Galvani discovered "animal electricity" produced by muscle.
1796	Edward Jenner established the procedure of vaccination.
1798	Thomas Malthus published *An Essay on the Principles of Population.*
	Erasmus Darwin observed adaptations and wrote about ecological relationships. He proposed the idea of sexual selection and said that all plants and animals are continually changing. He believed in the inheritance of acquired characteristics.

NINETEENTH CENTURY

1809	Jean-Baptiste Lamarck published *Zoological Philosophy,* describing his theory about the inheritance of acquired characteristics.
1815	William Smith published his map of the British Isles and later became lauded as the "father of geology."
	Georges Cuvier used his knowledge of comparative anatomy of living animals to make celebrated reconstructions of fossil animals from only a few of the fossil bones and proposed the theory of catastrophism.
1828	Friedrich Wöhler discovered organic molecules occurred in non-living materials.
1830	Charles Lyell published the *Principles of Geology.*
1831	The voyage of the *Beagle* began.
	Robert Brown discovered cell nuclei.
1836	Jön Jakob Berzelius named catalysis.
1838	Matthias Schleiden proposed that all plants are made of cells.
1839	Theodor Schwann generalized the cell theory to include animals.
1840s	Joseph Lister introduced antiseptic surgery to reduce infections.
1855	Rudolf Virchow used the cell theory to explain the effects of disease on the organs and tissues of the body.
1856	Louis Pasteur demonstrated that the processes in which beer, wine, and cheeses are made depended on the actions of microbes.
	First Neanderthal fossil discovered.
1857	Alfred Russell Wallace wrote to Darwin about his research in the East Indies and his theory.
1859	Charles Darwin's *Origin of Species by Natural Selection, or the Preservation of Favoured Races in the Struggle for Life* was published.
1850s	Robert Koch cultured and studied bacteria, associating specific organisms with tuberculosis and other diseases.
	Chromosomes named by Paul Ehrlich.
	Justus von Liebig began using chemical fertilizers in agriculture.
1865	Gregor Mendel published the results of his cross-breeding experiments.
1866	Ernst Haeckel coined the word *ecology.*
1870s	Walther Flemming discovered chromatin.
1882	Robert Koch identified the bacteria causing tuberculosis.
1896	Eugene Warming wrote *Ecology of Plants* and became the father of ecology.

TWENTIETH CENTURY

1900	Hugo De Vries, Carl Correns, and Erich von Tschermak-Seysenegg independently rediscovered Mendel's work; De Vries named mutations.
	Karl Landsteiner discovered the existence of different blood types.
1902	Walter Sutton proposed that chromosomes were the site of Mendel's factors.
1908	Godfrey Hardy and Wilhelm Weinberg independently proposed the Hardy-Weinberg principle of equilibrium.
	Archibald Garrod proposed the inborn errors of metabolism concept.
1910	Thomas Hunt Morgan used *Drosophila melanogaster* to show that genes are on chromosomes, and if genes are very near each other on the same chromosome, they tend to be inherited together.
1910s	Vitamins discovered and named.
1912	The most famous scientific fraud in history, Piltdown Man, was discovered in England.
1915	Alfred Wegener published the theory of continental drift.
1917	Joseph Grinnell developed the niche concept.
1920	Frederick Banting and Charles Best isolated insulin.
1921	Otto Loewi demonstrated that passage of a nerve impulse was chemical as well as electrical.
1924	Raymond Dart found the *Australopithecus* fossil called the Taung baby.
1927	Hermann Muller showed that X-rays cause heritable mutations.
1928	Alexander Fleming discovered penicillin.
	Frederick Griffith discovered transformation.
1920s	P. A. Levene found DNA was made of nucleotides and that each nucleotide was made of one of the nitrogenous bases adenine (A), thymine (T), cytosine (C), and guanine (G); a five-carbon sugar molecule; and a phosphate group.
1930s	First electron microscope built.
Early 1940s	Max Delbrück and Salvador Luria found that phages would infect bacteria and multiply within them until they filled the cells, which then burst open. They also found that the phages consisted entirely of DNA and protein.
1940	H. A. Krebs identified steps in the TCA cycle.
1941	George W. Beadle and Edwin L. Tatum published the results that led to the "one gene—one enzyme" hypothesis.
1944	Oswald Avery and colleagues Colin MacLeod and Maclyn McCarty showed transfer of DNA transformed *Streptococcus pneumoniae* from a harmless phenotype to a virulent phenotype.
	Alfred Hershey and Martha Chase showed phages' genetic material was DNA.

	Alfred Mirsky showed that in any species the somatic cells contain equal amounts of DNA, and the gametes contain only half as much.
	Erwin Chargaff found adenine and thymine were present in about equal proportions in DNA, and so were guanine and cytosine.
1940s	Melvin Calvin identified the reactions in photosynthesis.
1950	Barbara McClintock published proof of transposons in corn.
1951	Linus Pauling demonstrated the α-helical structure of proteins.
1952	Alfred Hershey and Martha Chase published data supporting the hypothesis that only DNA is required for T2 bacteriophage replication.
1953	James Watson and Francis Crick, together with Maurice Wilkins, proposed the double-helix structure of DNA, which was based on Rosalind Franklin's X-ray crystallography studies.
	Harold Urey and Stanley Miller performed the experiment that supported the Miller-Urey hypothesis.
1957	Arthur Kornberg discovered DNA polymerase.
1958	Matthew Meselson and Frank Stahl demonstrated that during replication, the two parental strands of DNA behave exactly as predicted by Watson and Crick's model.
1961	Marshall Nirenberg and Heinrich Matthaei demonstrated that the triplet UUU codes for the amino acid phenylalanine, the first step in deciphering the genetic code.
	Sydney Brenner, Francis Jacob, and Matthew Meselson demonstrated that ribosomes are the site of protein synthesis and confirmed the existence of messenger RNA.
1966	Marshall Nirenberg, Severo Ochoa, and H. Gobind Khorana deciphered the genetic code.
Late 1960s	A modified form of Alfred Wegener's theory (plate tectonics) finally became accepted by scientists.
1970	Howard Temin and David Baltimore independently discovered reverse transcriptase in RNA viruses.
1972	Paul Berg constructed a recombinant DNA molecule composed of viral and bacterial DNA sequences.
	Niles Eldredge and Stephen Jay Gould proposed the idea of punctuated equilibrium.
1974	*A. afarensis* (Lucy) was discovered in eastern Africa by Richard Johanson and his colleagues.
1977	Walter Gilbert and Fred Sanger independently developed methods for determining the exact nucleotide sequence of DNA.
1981	The first recombinant protein, human insulin, was marketed for the treatment of diabetes.
1984	Meteorite ALH184001 from Mars found in Antarctica.
1988	Kary Mullis developed the polymerase chain reaction (PCR) for amplifying large amounts of a specific DNA fragment from very small source amounts.

1989	Francis L. Collins found the gene responsible for cystic fibrosis.
1980s	Mice first successfully cloned.
	Genetically modified organisms appeared in agriculture.
1990	The Human Genome Project began.
	Gene therapy began.
1992	The entire 315,000-base nucleotide sequence of one of the chromosomes of the yeast *S. cerevisiae* was published.
1994	Yohannes Haile-Selassie found *Ardipithecus* fossil in Ethiopia.
1995	Craig Venter and others reported the first complete genome sequence of the bacteria *Haemophilus influenza.*
1996	Dolly the (cloned) lamb was born.
	Ozone layer began to recover.
1999	The first human chromosome (chromosome 22) was sequenced by scientists in the Human Genome Project.

TWENTY-FIRST CENTURY

2000	The Human Genome Project completed a "working draft" DNA sequence of the entire human genome.
2001	Two competing teams published complete maps of the human genome.
2003	Tim White and colleagues discovered a 160,000-year-old fossil of a modern human, supporting the Out of Africa theory.

Glossary

abiotic environment The nonliving parts of the environment.

absorbed Taken up. Products of digestion are absorbed through the intestinal wall into the blood vessels that surround the intestine.

abyssal zone The deepest part of the ocean, where it is very cold and dark.

accommodation Changes in the eye (especially in the lens shape) that allow seeing at different distances.

acid precipitation Precipitation formed when pollutants react with water in the air. Sulfur dioxide and oxides of nitrogen are the usual pollutants that form acid precipitation.

acidic Having a low (0–7) pH.

acidophiles Bacteria that live in very acidic areas, such as volcanic pools and hot vents on the sea floor, where the pH may be less than 1.

action potential During the passage of a neural impulse, sodium ions can flow into the cell, making the inside temporarily positive relative to the outside. This change is called an action potential.

active transport The movement of a substance against a concentration gradient, requiring the input of energy.

adapted Modified in a way that makes something fit for its surroundings. Organisms become adapted as a result of natural selection or may be preadapted for a change in environmental conditions.

adaptive convergence The process of different kinds of organisms becoming more alike as a result of selection by the same environment.

adaptive radiation The process of change during dispersal into new environments. Eventually different populations of a species may change in many different ways as they disperse, until many new species appear.

adenosine $C_5H_5N_5$, a purine found in nucleic acids that always pairs with thymine or uracil.

adenine triphosphate See ATP.

adrenal cortex The outer part of the adrenal gland, which makes corticotropic hormones.

adrenal medulla The inner part of the adrenal gland, which makes adrenaline.

advanced Recently evolved, usually in comparison to a more primitive group.

aerate Add air.

Agnatha The vertebrate class that includes jawless fishes.

albinism Lack of pigmentation.

algae A group of protoctists. Green algae are thought to be the ancestors of plants.

alkaline Having a high pH (7–14); basic.

alkaliphiles Bacteria that live in environments having a very high pH (more than 11).

allegory A statement about human existence told in the form of fiction, with symbolic characters.

alleles Different forms of the same gene.

allergen Proteins that can cause an allergic reaction in some people.

alluvial fans Fan-shaped deposits of sand or gravel left by streams, especially during floods. Common in deserts.

alpha-helix A right-handed coil found in proteins and most DNA.

amino acids The small units of which proteins are made, having the general formula NH_2–RCH–COOH. R stands for a group of atoms that are different in different amino acids.

ammonites Mollusks, now extinct but widespread in the Mesozic, that had flat spiral shells divided into chambers interiorly.

amniote Having eggs in which the embryo develops within a fluid environment enclosed by a sac, the amnion.

Amphibia The class of vertebrates that includes frogs and salamanders.

amygdala An almond-shaped structure in the temporal lobe of the cerebrum. It is involved in depression and other mental conditions.

anaphase The third stage in mitosis, which begins with the centromeres dividing. They are pulled toward the centrioles by the microtubules attached to them, with one member of each chromosome pair moving toward one end of the cell, and the other member moving toward the other end.

androgens The collective name for male sex hormones.

angiosperms Flowering plants, ranging from grasses to flowering shrubs to trees.

animals Organisms that can move spontaneously. They all are many-celled, and they are heterotrophic. All are in the kingdom Animalia.

ankylosaurids Dinosaurs that had tail clubs.

Annelida An animal phylum made up of worms with segmented bodies.

anther The tip of a stamen, one of a flower's male reproductive structures. Pollen grains are shed from the anther.

Anthophyta The plant phylum that includes angiosperms (flowering plants).

anthrax An infectious disease of cattle and sheep, caused by *Bacillus anthracis,* that can be spread to humans. In humans, it causes lesions of the skin and lungs.

anthropoids Advanced primates that more closely resemble humans than any other animals do. Comprising monkeys, great apes, and humans, anthropoids have larger brains and flatter faces than prosimians.

antibodies Proteins made by lymphocytes of the immune system. Antibodies are Y-shaped molecules; the ends of the antibody "arms" connect with antigens, taking them out of circulation or making it impossible for microbes to make toxins.

anticodon The point of attachment on a tRNA molecule that binds to a codon (triplet) on mRNA, making it possible for the tRNA to add an amino acid to a growing protein chain.

antigens Foreign proteins to which the immune system produces antibodies.

Archaea One kingdom in the three-kingdom classification system.

Archaean The period when life first appeared on Earth, 3.8–2.5 billion years ago.

archaeans Bacteria that are the oldest organisms on the Earth, having adaptations needed for surviving extreme conditions about 3.8 billion years ago.

archaic *Homo sapiens* Very early *Homo sapiens.* At one time Neanderthals were included in the group.

arithmetic growth Growth by addition of a constant amount at each step, in contrast to geometric growth.

arteriosclerosis Hardening and thickening of the arteries.

Arthropoda The animal phylum that includes insects and crustaceans. Arthropods have exoskeletons and jointed appendages.

artificial selection Selection of a plant or animal for breeding on the basis of desirable phenotypic characteristics.

Ascomycota One of the major groups of fungi, including yeasts and truffles. Most members of this group form sexual spores in little sacs.

asexual reproduction Reproduction by fission, budding, or other method that involves only one parent and usually results in offspring that are identical to the parent. No gametes are involved.

assortative mating Mating in which mates are preferentially chosen because of some genetic characteristic (sexual selection).

atom The smallest part of an element.

ATP (adenosine triphosphate) A three-phosphate molecule in whose phosphate bonds energy is stored at each turn of the TCA cycle and that can later be broken down for releasing energy.

autotroph Organism that makes its own food by chemosynthesis or photosynthesis.

Aves The chordate class that includes the birds. Birds are feathered, endothermic, and have many adaptations for flight.

axial skeleton The skeleton of the trunk.

bacteria The oldest forms of life. One kingdom (sometimes called Monera) in the five-kingdom classification system. Comprises archaebacteria and eubacteria. Prokaryotic; that is, bacterial DNA is not enclosed in nuclear membranes.

bacterial flora Normal bacteria found in or on the human gut or skin.

bacteriophages Viruses that feed on bacteria. Important in the early research on DNA.

Bactrian camel Type of camel found in Asia.

baleen Plates in the mouths of whalebone whales that filter out large numbers of very small organisms in marine plankton.

barophiles Bacteria that can withstand high pressure.

basal ganglia Masses of gray matter deep in the cerebrum.

basic High in pH (7–14); alkaline.

Basidiomycota A group of fungi including mushrooms, puffballs, and others. Most form millions of minute sexual spores.

behavior Way of acting. Useful adaptations may include behaviors.

benthic zone Oceanic zone below the pelagic zone, where the water is colder and darker than in the pelagic zone.

benzene A hydrocarbon and dangerous carcinogen, used as a solvent in manufacturing processes, such as the production of paints.

bilaterally symmetrical Having similar right and left halves. Only one plane can divide a bilaterally symmetrical organism into the two similar parts.

bile acids Acids produced by the liver and used in the intestine for emulsifying fats.

binomial nomenclature The method Linnaeus established for giving a unique name to each species. Each species has a unique two-part name (such as *Homo sapiens*) corresponding to its genus and species.

biodiversity The diversity of organisms and ecosystems in the natural world.

biological control Using natural controls, rather than pesticides, to control pests in agriculture and gardening. Predators may be introduced, or sterile males of the pest species may be released to prevent some of the population's growth.

biological fixation Conversion by organisms of the atmospheric form of an element or compound to a solid form, such as the conversion of nitrogen to nitrate or the conversion of carbon dioxide to glucose.

biome A very large ecosystem, such as the tropical forest biome.

biotechnology The manipulation of organisms (which may include genetic engineering) to produce useful products, such as pharmaceuticals.

biotic environment The organisms in an ecosystem.

bipedalism Walking on two legs. Only birds and human beings are bipedal.

bipolar disorder A mental condition marked by alternating periods of mania and depression.

birth control Prevention of birth, especially by interference with conception.

blastomeres Cells resulting from the first several divisions of a fertilized egg.

bolus Soft mass of food in the gut.

brachiopods Lampshells and other bivalve mollusks that use their arms to sweep food organisms into their shells.

Bryophyta Mosses and liverworts.

bryozoans A group of aquatic invertebrates that form mosslike colonies by budding.

bulbourethral glands Glands that add fluid to semen as ejaculation occurs.

Burgess shale Rock layer in the Canadian Rockies that contains the oldest animal fossils (about 670 million years old). Much evidence about the Cambrian lies in the Burgess Shale.

Calvin cycle A set of reactions during the second phase of photosynthesis in which carbon dioxide is reduced to carbohydrate. As each cycle begins, one molecule of carbon dioxide enters the cycle and is reduced; after six cycles, one molecule of glucose has been produced. The energy for the reactions comes from ATP.

canopy Umbrellalike forest roof formed by leaves of the tallest trees.

capillaries The smallest blood vessels, formed by branching of arterioles and venules. Networks of capillaries form the connections between arterial and venous blood vessels.

capsule Any sac enclosing fluid or tissue. A capsule may be tough, as in some plants, or membranous.

carnivores Animals that eat other animals.

carpel Vaselike structure containing a flower's ovary, in which an egg is formed. Located near the center of the flower.

carrier protein Protein molecule, embedded in a plasma membrane, that can transport polar molecules and ions across the membrane.

carrying capacity (K) Maximum population size that can be supported by the environment.

castings Excrement of earthworms, which enriches soil.

CAT (computerized axial tomography) A variation of linear tomography. CAT scanning uses no film; the results of many X-rays are recorded as patterns of electrical impulses by a radiation detector. Data from thousands of points are inte-grated and converted into mathematical information about the density of tissues at all the points.

catalysis Increase in rate of a chemical reaction, brought about by substances that remain unchanged by the reaction.

catalysts Chemical substances that make chemical reactions possible but do not participate in the reactions themselves.

catastrophism Cuvier's theory that each of a series of catastrophes had made old forms extinct and allowed the creation of new life.

cell wall The structure surrounding a plasma membrane and providing protection and support for the cell. Found only in nonanimal cells.

Cenozoic The era that began 65 million years ago and continues today. Its name is based on the Greek word *coenos* (recent).

central dogma of genetics In a one-way process, the deoxyribonucleic acid (DNA) of the nucleus causes the synthesis of ribonucleic acid (RNA), which causes the synthesis of protein.

central sulcus The deep crevice that crosses the brain from left to right.

centrioles Cellular structures formed of microtubules. Each cell in an animal contains two centrioles. During cell division they duplicate and move to the ends of the spindles, and the chromosomes move toward them.

centromere The point at which two chromatids are attached to each other.

cerebrum In the human brain, the largest portion. Learning, memory, and other sensory and cognitive functions are located in it.

channels Tunnels.

characteristic Property. Often used as a synonym for *trait*. In Mendelian genetics, characteristic is a kind of property, such as eye color, and trait is the specific form of it, such as red or white.

charophycean Green algae shown to be more closely related to land plants than other green algae are.

chemical bonds Bonds between the atoms in chemical compounds. Energy can be stored in some chemical bonds for later use.

chemical reaction Reaction between two or more atoms or molecules in which chemical bonds are made or broken.

chemoreceptors Structures that are sensitive to the presence of chemicals.

chemosynthetic bacteria Bacteria near hydrothermal vents on the ocean floor that use the energy in hydrogen sulfide and other compounds to manufacture food.

chlorophyll The green pigment found in chloroplasts in cells of photosynthesizing organisms. Chlorophyll is the receptor of light energy from the sun.

chloroplast Structure that contains chlorophyll in cells of photosynthesizing organisms. Contains DNA. Probably these organelles began as prokaryotes that were absorbed by larger cells and developed a symbiotic relationship with them.

choanocytes Food-retaining cells in a sponge's walls. Choanocytes resemble choanoflagellates, so sponges probably arose from a flagellate ancestor.

choanoflagellates Protoctists that swim by lashing their flagella. Probably the ancestors of sponges.

cholesterol A steroid alcohol, $C_{27}H_{45}OH$, important in many metabolic processes. High levels in the blood have been linked to increased risk of cardiovascular disorders, such as heart attack and stroke.

Chondrichthyes Sharks and other cartilaginous fish.

Chordata The animal phylum comprising humans and other animals having notochords.

chromatids The two identical strands of chromosomes that are attached to each other at a point called the centromere.

chromatin Material made of nuclear DNA and associated proteins that condenses and becomes visible under a light microscope just before mitosis begins.

chromatophores Internal membrane systems in autotrophic bacteria that contain pigments for photosynthesis.

chromosomes Large, dark-staining double-stranded structures in dividing cell nuclei. They are made of DNA and protein in eukaryotes and DNA alone in prokaryotes. The genes are on chromosomes.

chylomicrons Protein-coated droplets in lymph, in which fatty acids absorbed from the intestine travel to the point in the chest where it empties into the circulatory system.

cilia Short flagella. They may be used as "oars" by swimming prokaryotes, or as structures for moving things across surfaces by more advanced organisms.

circadian rhythms Approximately 24-hour cycles of growth or activity.

clay One of the three chief inorganic ingredients of soil. Made of very fine particles of aluminum silicates, clay causes soil to retain water.

climate The long-term weather conditions in an area, especially temperature and precipitation.

climax community The community once thought to end a rigid order of succession.

clone A population of genetically identical cells or organisms that are derived originally from a single original cell or organism by asexual methods. Or an individual that was grown from a single body cell of its parent and thus is genetically identical to it.

Cnidaria The animal phylum containing *Hydra*. Cnidarians have jellylike bodies and two cell layers; many have tentacles; many are radially symmetrical.

cochlea A structure in the inner ear containing hairs that vibrate when sound waves reach them. The vibrations are translated into nerve impulses that go to the brain through auditory cranial nerves.

codominance Inheritance pattern in which heterozygotes show the effects of both the dominant and recessive alleles, such as in blood types.

codon A triplet of mRNA carrying the code for one amino acid.

coelom A cavity in the mesoderm that is characteristic of all advanced animals.

coenzymes Chemicals that are needed for the functioning of enzymes. Some vitamins—notably the B vitamins—act as coenzymes.

coevolution The evolution of two or more species that adapt to each other, such as insects and flowering plants.

collagen A protein found in bone. Fibers of collagen give the durable skeleton its flexibility.

commensalism A symbiotic relationship in which one species benefits, and the other is neither helped nor harmed.

community The various populations in an area.

compound A chemical formed of two or more elements that are chemically combined.

compound microscope A microscope having at least two lenses.

computerized axial tomography See CAT.

cones In plants, seed-bearing structures. In vertebrate animals, receptors that transmit information about visual details and color to the brain through the optic nerve.

Coniferophyta The plant phylum that includes pines and other conifers.

conjugation Union of two unicellular organisms, in which genetic material flows between them.

continental drift Movement of continents. Now called plate tectonics, Wegener's original theory was called continental drift.

Copernican theory The theory of Nicolaus Copernicus (1473–1543) that the Earth revolves around the sun rather than being the center of the universe.

coronary arteries Arteries that lead to the heart muscle itself.

crania Skulls without the lower jaws.

cranial nerves The 12 nerves connecting directly to the brain.

creationists People who believe the Earth was created literally as described in Genesis and reject scientific theories of evolution.

crinoids Sea lilies, primitive animals that became extinct after the Permo-Triassic crisis.

Crohn's disease A disease that causes painful inflammation and ulcers and may be caused by a defective gene.

cross-pollination Fertilization of one individual plant by the pollen from another.

cross-section A cut across an organism, or part of one, made at right angles to its long axis.

crossing over The exchange of DNA from either chromosome with that from the other, as the result of members of each homologous pair moving close together during prophase. Thus, genes from the father's side of the family may be exchanged with those from the mother's side, and the individual chromosomes are no longer maternal or paternal.

crust Earth's hard surface.

culture To grow bacteria in the laboratory by supplying the appropriate temperature, moisture, and nutrients. The cultivated bacteria are also called a culture.

cuticle In invertebrate animals, a body covering made of chitin and protein that provides the animals with a tough exoskeleton for support and protection. In plants, a waxy coating on leaves that limits water loss.

cyanobacteria A major group of photosynthetic bacteria. Cyanobacteria live on root nodules of legumes and convert N_2 to ammonium ions (NH^{4+}) or nitrate ions (NO^{3-}), which can be taken up by the plants' roots and used for making protein. Formerly called blue-green algae.

Cycadophyta Cycads.

cytology The study of cells.

cytoplasm All of the cellular material outside the nucleus.

cytosine $C_4H_5N_3O$, a pyrimidine found in nucleic acids that always pairs with guanine.

cytoskeleton A network of filamentous proteins that gives the gel-like cytoplasm some support.

Daily Value The daily allowance of a nutrient or vitamin that is recommended by the U.S. Food and Drug Administration.

degenerate Having more than one triplet (codon) for an amino acid.

dehydrated The condition resulting from losing too much water or body fluids.

deletion Loss of part of a chromosome during cell division.

dementia Mental deterioration.

denitrification Conversion of nitrogen compounds in soil or water to N_2 by soil bacteria.

density Size divided by the area, as in population density.

detoxification Breakdown of toxic substances.

detritus Excreta and other wastes.

dialysis Use of an artificial kidney to remove wastes from the body.

dichotomous key A scheme in which you begin with a general description of an organism to be identified and proceed through a series of either/or choices until you reach the organism's name.

dicots Dicotyledonous plants. They have two seed leaves, branching leaf veins, and other shared characteristics.

differentiate To go from a general, unspecialized state to a specific, specialized state.

digestion Breakdown of foods to simple, absorbable units.

digestive system System where digestion occurs, made up of the stomach, pancreas, and other organs.

diploid number Number of chromosomes in body cells ($2n$ number); for example, 46 in humans.

disaccharide Sugar consisting of two simple sugars linked by a chemical bond.

disassortative mating Mating in which the partners differ a great deal genetically.

DNA Deoxyribonucleic acid, the substance of heredity. It has a helical structure made of two linked strands. The backbone of each strand is a chain of ribose (a five-carbon sugar) and phosphate groups having nitrogenous bases (A, T, C, and G) attached; and the two strands are linked by hydrogen bonds between the base pairs (A–T and C–G).

DNA fingerprinting Using a person's DNA to prepare a unique pattern for comparison with the patterns from other people.

dominant An allele that determines what a genetic characteristic will be, even if the other allele of the pair is different.

Drosophila Fruit flies, organisms used often in genetic studies.

duodenum The first part of the small intestine, where most digestion occurs.

Echinodermata The animal phylum that includes starfish and other spiny-skinned invertebrates.

ecology The study of interrelationships among living things and their environment.

ecosystem A community of organisms and their abiotic environment.

ecotourism Traveling for the purpose of studying the environment, and respecting the environment during travel.

ectoderm In all animals that have three cell layers, the layer that gives rise to the skin and nervous tissue.

elements The simplest chemicals, such as carbon (C) and hydrogen (H). A single atom is the smallest part of an element.

embryo The early developmental stage of a vertebrate animal, formed by divisions of a fertilized egg. In humans, at eight weeks after conception the embryo becomes a fetus.

emigration Movement out of an area.

endangered Threatened with extinction.

endocrine system A system made up of glands. Unlike other systems, it is scattered through the body.

endoderm In all animals that have three cell layers, the layer that gives rise to the linings of all tracts, such as the digestive and urinary tracts.

endometrium The lining of the uterus, where a fertilized egg is implanted and develops. If there is no fertilization, the endometrium is shed (menstruation).

endoplasmic reticulum See ER.

endorphin A peptide, produced in the hypothalamus, that softens pain perception.

endoskeletons The bones that protect internal organs in vertebrates. They provide no external protection, as the exoskeletons of invertebrates do.

entropy The physical principle that all matter tends to become less organized. Because of entropy, maintaining organized living systems requires continual energy.

environment Surroundings.

enzymes Protein catalysts that make breakdown and synthesis of molecules possible. Often used in biotechnology.

epididymis A coiled part of the vas deferens, where sperm are temporarily stored and mobilized.

epiphytes Plants that are not parasites but grow on the branches of other plants.

epithelial Surface layer of cells on skin or lining a body cavity.

ER (endoplasmic reticulum) A network of sacs, channels, and tubes folded in the cytoplasm.

estivating Becoming dormant (usually in the dry season or summer).

Eubacteria All bacteria except the archaeans. May be free-living or parasitic.

eukaryotes Organisms in which the cells have a centrally located nucleus containing the cell's DNA. The prokaryotes and eukaryotes are the two great divisions of all living things.

eutrophication Overgrowth of plant life in fresh water, with resulting decay, as a result of too many nutrients enriching the water.

evolution Permanent, heritable changes in populations of organisms over time.

excretory system The system that helps keep the ranges of salts and water in the tissue fluid normal and eliminates cellular waste products from the body.

exons The coding regions of genes.

exoskeletons The external cuticles or shells that protect the bodies of invertebrates.

exponential growth The pattern of growth of biological populations, which has an S-shaped curve. The number of individuals increases slowly at first, then rises quickly until it reaches the carrying capacity. If it is graphed with the exponents of size on the y axis instead, a straight line will result.

expressed Discernible, as in an organism's phenotype.

expression The discernible effect of a gene.

extinction Permanent loss of a species.

extracellular digestion Digestion in the body cavity of cnidarians.

extremophiles Archaebacteria; adapted to extreme environmental conditions.

facilitated diffusion Diffusion across a membrane in which the molecules do not dissolve in the phospholipid layer of the membrane. Instead, proteins in the membrane bind the molecules and help them move across it.

falsification of hypotheses According to philosopher Karl Popper, the way a scientist approaches the truth: by gathering observations about the natural world, then forming a tentative hypothesis to explain the observations. Further observations or experiments may support the hypothesis indefinitely; if even one reliable observation or controlled experiment falsifies the hypothesis, it must be rejected or revised.

fat Solid form of the class of chemicals called lipids, which include triglycerides, fatty acids, phospholipids, and sterols.

fatty acid A long hydrocarbon chain ending in –COOH, found in lipids. End product of digestion of fats.

feedback Information about the level of a molecule that brings about the production of more of the molecule (positive feedback) or a lessening of its production (negative feedback).

fermentation Anaerobic respiration, which produces alcohol, acids, and carbon dioxide from carbohydrates.

fetus A developing but unborn vertebrate animal that has passed through the first stages of development. In humans, at about eight weeks after conception the embryo becomes a fetus.

fiber Threadlike strand; there are plant fibers (that make up roughage in foods), muscle fibers, nerve fibers, and other fibers in plants and animals.

filament Lower part of a stamen, ending in an anther.

filters Passes through a sieve or membrane, removing particles.

fission The simplest type of asexual reproduction. The splitting of one organism into two, each of which is identical to the parent organism.

five-kingdom system The classification system most in use today. It includes the kingdoms Plantae, Animalia, Fungi, Protoctista, and Bacteria.

flagella Whiplike tails that can be used for swimming, found in bacteria and protoctists.

flavors Combinations of tastes and odors.

flowers The sex organs of angiosperms.

follicle Part of an ovary in which an egg starts to mature.

fossil fuels Fuels made by the gradual conversion of plant and animal remains by temperature and pressure. Natural gas, petroleum.

fossils Evidence of past organisms.

fractional distillation Separation of crude oil into substances, such as gasoline, lubricating oil, and paraffin, by heating so that each vaporizes or condenses at a specific temperature.

frame shift Mutation caused by a DNA sequence being misread during transcription or translation. The resulting mRNA would code for a completely different sequence of amino acids. Sometimes a frame shift results in an abnormal protein that causes a malformation or death; or no protein at all may be formed.

fronds Large, divided leaves; those on ferns, for example.

fructose A simple sugar. Combines with glucose to form sucrose (table sugar).

Fungi One of the five kingdoms of organisms. Includes molds, lichens, yeasts, and others.

GABA (gamma-aminobutyric acid) A neurotransmitter found in many synapses.

game theory The theory of strategies that will work in different situations or in response to different opponents. Evolutionary biologists often find themselves explaining animals' behaviors in terms of game theory.

gametes Sperm and egg cells; germ cells.

gamma-aminobutyric acid See GABA.

gastric juices Acids that digest foods, found in the stomach.

gene The unit of inheritance. The amount of DNA providing enough information for the cell to assemble one protein from amino acids in the cytoplasm. Sometimes it is defined in terms of providing information for one RNA segment.

gene flow The movement of gene-carrying individuals from one population to another, which tends to make different populations genetically similar.

genetic code The sequence of bases in mRNA. Each base can be decoded as an amino acid.

genetic distance Distance between populations as indicated by the difference between their frequencies of a given allele.

genetic drift A change in gene frequency in a small population, made possible because chance alone can have a large effect on a small population.

genetic engineering Making artificial changes in DNA.

genetic recombination The rearrangement of genes and chromosomes that can occur during meiosis. This is the source of much variation, even within families.

genetically modified organisms See GMOs.

genome The total DNA of an organism or a species.

genomics Study of sequence and structure of the genome.

Geographic Information System See GIS.

geologic era A major division of time, lasting as long as hundreds of millions of years, based partly on geology.

geometric growth Growth by multiplication at each step, contrasted with arithmetic growth. In populations it leads gradually to a point where the number "explodes" suddenly.

germ cells Gametes; as opposed to body cells.

germination Emergence of a growing plant embryo from a seed.

gill arches The bony arches separating a fish's gills.

Ginkgophyta Ginkgo plants.

GIS (Geographic Information System) A computer-equipped satellite system that can assemble, manipulate, and display nearly any information in relation to a region's geography.

glands Organs that secrete hormones or other substances.

glucocorticoids A group of steroids that affect the metabolism of foods and are anti-inflammatory. Often used as drugs for rheumatoid arthritis.

glycogen The form in which glucose is stored in animal tissues.

GMOs (genetically modified organisms) Animals or plants that are the products of genetic engineering.

Gnetophyta A plant phylum, the members of which bear seeds in cones.

Golgi complexes Cup-shaped cytoplasmic structures that form lysosomes and other membranous structures.

Gondwanaland A southern continent formed by the breakup of Pangaea, which gave rise to South America and Africa.

gray matter Neural tissue containing cell bodies of neurons. Found on the surface of the cerebrum and in the interior of the spinal cord.

green algae A group of protoctists, the ancestors of modern plants.

guanine A purine, $C_5H_5N_5O$, found in nucleic acids that always pairs with cytosine.

guard cells Cells that can open or close the stomata of plants as needed.

habitat Area where an organism lives.

half-life The time during which half of an original (parent) radioactive element breaks down to a daughter element.

halophiles Archaeans that can tolerate very salty environments.

haploid number Number of chromosomes in gametes ($1n$ number); for example, 23 in humans.

Hardy-Weinberg equilibrium principle The mathematical principle showing that generation after generation, if there is random mating in an infinitely large interbreeding population, and there is no selection, migration, or mutation, the proportions of alleles and genotypes remain the same.

headwaters The springs, snowmelts, or lakes that begin streams and rivers.

hemoglobin The pigment in blood that carries oxygen.

herbivorous Using plants for food.

heredity The transmission of characteristics to offspring.

hermaphrodite An individual plant or animal possessing both male and female sex organs.

heterotrophic Unable to manufacture its own food.

hibernate Become dormant in winter.

hierarchy A system in which closely related organisms are grouped together under a heading, closely related categories are grouped together under another heading, and so on up to one heading. In biological classification, the top heading is "living things."

hippocampus A small structure in the cerebrum that is shaped like a seahorse and is involved with memory.

holograms Three-dimensional images.

homeostasis Equilibrium.

hominid Prehumans and humans.

hominoid Hominids and pongids.

Homo The human genus.

homolog One of the members of a pair of chromosomes that appear identical. One carries genetic information from the maternal line (from the individual's mother), the other paternal (from the individual's father).

homologous Identical in appearance.

homozygous Having the same allele of a gene on both homologous chromosomes.

hornworts Aquatic bryophytes having a hornlike projection.

host Animal on which a parasite or virus depends.

Hox genes A group of genes that direct the process of assigning blastomeres to their addresses in the developing organism.

Huntington's disease An ultimately fatal genetic disorder that begins with uncontrollable body movements and mental deterioration, caused by the death of neurons in the brain's basal ganglia.

hybrid The offspring of parents of different breeds, genera, or species.

hydrogen bonds The bonds between the positive or negative poles of water molecules and other polar molecules or ions. Hydrogen bonds link the parallel chains of nucleotides in DNA.

hydrolysis The splitting apart of a compound, with the hydrogen and hydroxide ions in water being attached to the new attachment points.

hydrolytic Performing or aiding hydrolysis.

hydrothermal vents Openings in the midocean ridges at the bottom of the abyssal zone, from which hydrogen sulfide and other minerals bubble up into the water and provide energy for chemosynthetic bacteria.

hypothalamus Part of the brain that secretes hormones stored in the pituitary.

igneous rock Rock formed by the cooling and hardening of lava from volcanoes.

immigration Movement into an area.

immune reaction Antigen–antibody reactions occurring in response to infection; if they are completely successful, the infection is cured. The effects of an immune reaction may be very obvious, such as swelling, pus formation, or redness.

immune system The system that resists bacterial infection and other foreign proteins on three levels: first, by preventing their entering the tissues at all; second, by using scavenger cells to remove them from tissue fluid; and third, by producing antibodies that can tie up bacterial antigens chemically.

immunity Protection of the body against disease.

in vitro fertilization Procedure in which eggs are fertilized in the laboratory before being implanted in a woman's uterus.

inborn errors of metabolism Hereditary disorders in which the body can't perform some normal chemical process, resulting from abnormal genes coding for abnormal enzymes.

inbreeding Breeding between close relatives, which often leads to the accumulation of deleterious genes that may cause defects or predispose the individual to disease.

incomplete penetrance Some individuals in a population not showing any effects of an allele known to be present.

index fossils Fossils that indicate the age of the rock layer in which they are found.

infection The establishment of bacteria in a host.

initiator The first codon on the mRNA (usually AUG). The corresponding tRNA anticodon pairs with the initiator codon, bringing its attached amino acid along. (If the initiator is AUG, the amino acid is methionine.)

innate Inborn.

insulin A hormone produced by the islets of Langerhans in the pancreas. Insulin enables cells to use sugar.

interphase The resting stage in cells, when the cell is not undergoing mitosis.

intertidal zone The area that is covered at high tide and exposed at low tide. It may be a broad sandy beach or a steep rocky shore.

introns The noncoding regions of genes.

invertebrates Animals without spines.

ion channels Water-filled tunnels that open up in the membrane of a nerve cell, making the inside temporarily positive relative to the outside.

ionized Converted in part or in whole to ions.

isotopes Different forms of an element, differing in the number of neutrons and behaving differently.

jumping genes Transposons.

junk DNA Noncoding DNA in a chromosome.

karyotypes Photos of the chromosomes in a single cell of a particular organism.

labyrinthodonts Early amphibians that became dominant life forms and diverged into many species during the Carboniferous period.

lacteals Vessels leading from the villi lining the intestine to the lymphatic system.

lactose A sugar found in dairy foods.

lacunae Lakelike spaces.

larva The stage that follows the egg in invertebrates that undergo metamorphosis.

Laurasia A northern continent formed by the breakup of Pangaea, which gave rise to North America, Europe, and Asia.

LDL (low-density lipoprotein) A lipoprotein containing a high proportion of cholesterol and associated with increased risk for atherosclerosis.

legumes Plants having seeds in pods, such as peas and beans.

lichens Algae and fungi that live together mutualistically.

ligases Enzymes that paste DNA segments together.

litter Decayed layer of leaves and other organic matter.

locus Point on a chromosome where a gene (or allele) is located.

Lycophyta Club mosses.

lymph Tissue fluid that is within the lymphatic system. Lymph is also the plasma portion of blood.

lymphatic system A network of vessels that move lymph around the body and empty it into some veins in the chest.

lysosome Membranous sac that can fuse with a vacuole containing bacteria and digest the bacteria, using enzymes.

macroevolution The great overall picture of evolution, showing how entire organisms and groups have changed over time.

magnetic resonance imaging See MRI.

malaria A human disease carried by mosquitoes and caused by a sporozoan parasite. Malaria is marked by chills and fever.

malignant Cancerous; infiltrating other body parts.

Mammalia The class of vertebrates that includes the mammals.

mammals Animals that have hair and nurse their young.

mandibulates Arthropods having strong, biting jaws.

mantle A body covering that secretes a hard shell in many mollusks.

marker Chemical added to DNA or other molecule that can be traced for the purpose of discovering where the molecule has traveled.

medulla oblongata The most posterior part of the brain; the enlarged end of the spinal cord.

medusa In cnidarians, a bowl-shaped form. Or, the body may be tall and cylindrical (a form called a polyp).

meiosis A special form of cell division occurring during formation of the gametes, resulting in cells with the haploid number of chromosomes.

menstruation Monthly shedding of the uterine lining when fertilization does not occur.

meristem The plant tissue that grows and adds to the plant's size. It forms new cells throughout the plant's life by mitosis. Meristems are found at the outer ends of stems, roots, and buds. This type of meristem produces growth in length, or primary growth. Secondary growth is growth in diameter, which is brought about by meristem between xylem and phloem in the stem.

mesoderm In all animals that have three cell layers, the layer that gives rise to muscle and bone as well as various internal organs.

mesoglea In cnidarians, a substance that separates the two body layers. It contains contractile fibers and nerve cells but is not a true cell layer.

mesophiles Archaeans that are adapted to moderate environmental conditions.

Mesozoic The geologic era beginning 245 million years ago and ending 65 million years ago. Its name comes from the Greek word *mesos,* meaning middle.

metabolize Break down by metabolic processes.

metamorphosis Development through different stages from egg to adult (as in a butterfly, for example).

metaphase The second stage in mitosis, during which the chromosomes line up along the equator of the nucleus, in a plane that is perpendicular to the nucleus. Microtubules are attached to the centromeres.

meteorites Meteors that have reached the surface of Earth in solid form.

microbes Bacteria and other microorganisms.

microevolution Changes in DNA that are the microscopic basis of evolution.

microtubules Tubules in the cytoplasm from which are formed the mitotic spindle fibers and centrioles. They are attached to the centromeres during anaphase. Cilia and flagella are also made of microtubules.

midbrain Part of the brainstem, along with the medulla and pons. The brainstem controls vital body activities, such as breathing and heartbeat.

Miller-Urey hypothesis The hypothesis that if life arose during the atmospheric conditions of the Archean, it might be possible to reproduce the origin of amino acids, at least in the laboratory.

minerals Naturally occurring inorganic compounds. They cycle through the nonliving and living parts of mineral cycles. In nutrition, minerals are inorganic materials that are needed for formation and functioning of various body parts and compounds.

mini-satellites Repeated 15-nucleotide segments in human chromosomes. The number of repetitions and locations can vary greatly and are unique in every person, making it possible to construct DNA fingerprints of people.

mitochondria Sites of energy extraction from food in the cytoplasm of cells. Mitochondria have their own DNA (mitochondrial DNA) and are thought to have originated as free-living bacteria that were absorbed by cells.

mitochondrial DNA See mtDNA.

mitosis Cell division resulting in cells with the diploid number of chromosomes. Mitosis is not a simple splitting in two, but a complicated series of steps. The major steps are interphase, prophase, anaphase, metaphase, and telophase.

molecular genetics Genetic studies that focus on the DNA molecule and its chemical actions.

molecules The smallest part of a compound, formed by combining atoms.

Mollusca The phylum that includes clams and snails. Mollusks are soft-bodied and usually surrounded by a protective shell.

monoclonal antibodies Highly uniform antibodies against a specific antigen, prepared by combining tumor cells with the antibodies. The combined cells are called hybridomas.

monocots Monocotyledonous plants. They have one seed leaf, parallel leaf veins, and other shared characteristics.

monoculture Agriculture based on just one crop or plant variety.

monoglycerides The products of triglyceride digestion.

motile Able to move independently.

motor Pertaining to motion. Motor neurons of the somatic nervous system innervate the skeletal muscles, which an animal moves voluntarily.

motor cortex The part of the cerebral cortex that lies mainly anterior to the central sulcus. Each point on it is linked remotely to some motor effector in the body.

MRI (magnetic resonance imaging) A method for providing images of soft tissue without using X-rays. A strong magnet causes certain atomic nuclei to align with the magnetic field. Radio waves are then used to stimulate the magnetized nuclei in the patient's tissues; a computer picks up signals from the nuclei and converts them to visible images. MRI was originally called nuclear magnetic resonance (and sometimes still is), but the word *nuclear* was misinterpreted as referring to radiation, and so it was dropped.

mRNA (messenger RNA) The RNA that carries genetic information from the nuclear DNA to the cytoplasm.

mtDNA (mitochondrial DNA) DNA found in mitochondria. Unlike nuclear DNA, mitochondrial DNA is passed to offspring by the mother only, a characteristic that has been useful for tracing patterns of evolution.

multicellular Made of many cells.

multifactorial The form of inheritance in which characteristics are found in several forms, not just in two.

multiple alleles The various forms of a gene. Most genes have multiple alleles, but only two can be present at that gene's location on the pair of chromosomes.

mutation Any change in a gene's ordinary form, such as substitution of one base for another or deletion or insertion of a base.

myth A story about a people's history, told to emphasize a world view.

natural selection Charles Darwin's proposed mechanism for the origin of species. As a result of natural selection by the environment, individual organisms continue to survive and reproduce, or they die. Selection acts on all the anatomical, physiological, and behavioral characteristics of an organism, so the genes determining them are selected for or against in the process.

Neanderthals A mysterious group of hominids whose remains are 28,000 to 40,000 years old. The classic Neanderthals—those first described—are classified as *H. neanderthalensis,* a separate species. Some closely related fossils, though, are given the name *H. sapiens* with the subspecies *neanderthalensis.*

Nematoda A phylum of round-bodied worms; many are transparent.

nephridia Excretory structures found in annelids.

nerve impulses Electrical impulses traveling through neurons.

nerve nets The simplest nervous systems that connect all parts of an animal. Found in many cnidarians.

nervous system System of nerves and brain, which controls sensations, movements and activities, and thought and language.

neuromuscular junction The place where a nerve impulse reaches a muscle and calcium is released.

neurons Nerve cells.

neurosecretory Carrying electrical signals and secreting hormones into the bloodstream.

neurotransmitter A chemical that is released into the synapse between the axon of one neuron and the dendrites of another, making it possible for a nerve impulse to travel between neurons.

niche Way of life, including feeding relationships.

nitric acid An acid formed by nitrogen and water, often found in rain or snow.

nitrogen fixation The conversion of nitrogen to nitrogen compounds by soil organisms.

nitrous acid An acid formed by nitrogen and water, often found in rain or snow.

noncoding Segments of the chromosomes, sometimes called introns or junk DNA, with no known function.

nondirectional variation Genetic variation occurring in many directions.

nondisjunction The result of chromosomes and chromatids not separating normally during mitosis or meiosis, so that the daughter cells receive too many or too few chromosomes.

normal distribution The theoretical distribution of characteristics in a population that is shown as a bell-shaped curve.

normal flora Normal microbiota.

normal microbiota Bacteria normally living on the skin or in the gut.

notochord A long, stiff rod extending along the body in chordate embryos that supports the body. It is retained throughout life In a few primitive chordates, but in most groups it is eventually replaced by cartilage or bone.

nuclear envelope Membrane enclosing a nucleus.

nucleolus Dense spherical structure inside the nucleus, where the DNA is copied to mRNA.

nucleotides Chemical units of DNA. A nucleotide is made of one of the nitrogenous bases adenine (A), thymine (T), cytosine (C), and guanine (G); a five-carbon sugar molecule; and a phosphate group.

nucleus Interior compartment of a cell containing most of the cell's DNA.

nutrients Substances that can be absorbed and used by cells for energy and materials. The three main classes are proteins, fats, and carbohydrates.

nutrition Provision of nutrients.

nutritional genomics Use of genetic information to determine individual nutritional requirements.

oil Lipid in a liquid form.

oligochaetes The group of annelids that includes the earthworms.

omnivorous Eating both plant and animal foods.

oncogenes Genes that are activated by some retroviruses, producing proteins that control cell growth and leading to the uncontrolled cell growth and division characteristic of cancer.

organisms Living things.

orgasm The climax of sexual intercourse. It includes involuntary muscle contractions and the release of seminal fluid by the man.

osmosis Diffusion occurring across a membrane.

osteoporosis A condition in which the bone loses so much density that it becomes porous and brittle.

Out of Africa hypothesis The hypothesis that modern humans are all descended from a single ancestral African group of humans who lived about 170,000 years ago and that their descendants continued to live on that continent until around 50,000 years ago. Then they began emigrating, eventually spreading around most of the planet.

ovaries Female sex organs, in which eggs are formed.

oviducts Passages through which eggs travel from the ovaries to the uterus.

ovule Structure at the center of a flowering plant's ovary, within which the egg matures.

pacemaker Heart tissue that controls the strength and rate of the heartbeat.

paleontologists Scientists who study and interpret fossils.

Paleozoic The geologic era beginning 544 million years ago and ending 245 million years ago. Its name comes from the Greek word *palai,* meaning ancient.

panacea A cure-all.

Pangaea A supercontinent formed by small land masses joining together by the beginning of the Mesozoic.

pangenesis Process hypothesized by Darwin but later discredited. He thought an organism's body cells might accumulate small particles during growth and throughout the life span. The particles could then pass from body cells to germ cells (sperm and ova), where they could pass some changes on to the organism's offspring.

panspermia hypothesis The hypothesis that life on Earth did not evolve here but was seeded by spores or microbes that arrived on meteorites.

paradigm Theoretical framework leading to new hypotheses and experiments.

parasite An organism that depends on another organism (its host) for nutrition.

passive diffusion The diffusion of compounds from an area where they are more concentrated to an area where they are less concentrated without any assistance.

pathogens Disease-causing organisms.

pectins Compounds in cell walls that bind cells together; they yield the gels found in fruit jellies.

pelagic zone The cold, open ocean.

peptide bond The bond between carbon and nitrogen in a peptide linkage (CONH, or CO–HN). Long chains of 50 or more amino acids, linked by peptide bonds, make up the proteins.

peristalsis Muscular movements that move food through the digestive tract.

peritoneal fluid Fluid in the coelom, which is lined with a membrane, the peritoneum.

permeable Penetrable.

Permo-Triassic crisis A major extinction at the end of the Paleozoic era.

PET (positron emission tomography) A diagnostic scanning tool that uses radioactive compounds to generate gamma rays that can be detected to produce images with a computer; used for studying heart and brain functions.

pH Acidity or alkalinity. The pH scale ranges from 1 to 14, with 1 being very acidic, 7 neutral, and 14 very alkaline.

phages See bacteriophages.

phagocytosis The process of engulfing and digesting.

pharmacogenomics The use of an individual's genetic information to determine how they will respond to drugs.

pharynx A muscular tube at the back of the throat that leads both to the esophagus and the bronchi.

phenotype The discernible effects of the genotype.

pheromone Chemical substance emitted by an organism that affects the behavior of another organism of the species. Many pheromones are sexual attractants.

phloem Conducting vessels through which the food made during photosynthesis passes from chloroplast-containing cells to the rest of a plant.

phospholipids Lipids that contain phosphorus atoms. Phospholipids are found in cell membranes.

photoperiodism Regulation of an organism's physiology by light.

phylogenetic tree An evolutionary family tree, showing how a group of ancestral organisms has branched over time.

phylogeny Evolutionary relationships among groups of organisms.

pili Hairlike structures on the surface of bacterial cells.

pineal gland Gland inside the brain that produces the hormone melatonin and affects circadian cycles.

pistil The part of a flower that contains an egg.

pituitary gland The master gland of the body, next to and controlled by the hypothalamus. Secretes a variety of hormones that affect reproduction and other activities.

placebo A "sugar pill" or other noneffective drug. A counterfeit medication, used at one time by physicians to make patients feel better when no effective medicine was available. Placebos are used today in controlled studies of new drugs.

placenta Tissue in the uterus through which an embryo is nourished, containing blood vessels from both the mother and the embryo.

plains In the United States, the western portion of the grassland biome, east of the Rocky Mountains.

plants Nonmotile, multicellular, autotrophic organisms. Cells surrounded by cell walls; food stored as starch; chloroplasts in cells; autotrophic; develop from an embryo. All are in the kingdom Plantae.

plaques Deposits that narrow the arteries, formed by cholesterol and other substances associated with fat.

plasma The fluid portion of blood.

plasmid A DNA fragment that is not part of the chromosomes. Plasmids are more common in bacteria than in eukaryotes.

plate tectonics The theory that continents move because they are the higher portions of huge crustal plates floating on a hot, semi-liquid rock layer called the asthenosphere. Plate tectonics replaced Wegener's theory of continental drift.

platelets Minute fragments of blood cells that are important in forming clots.

Platyhelminthes The animal phylum that includes liver flukes and planarians. These animals are flat, with bilaterally symmetrical bodies; have three cell layers; many are parasitic.

pleiotropy Inheritance in which one gene can affect many parts of the phenotype that appear unrelated.

pollen tube A tube that grows down through a stigma to the ovule, where fertilization occurs.

pollinators Animals and other agents (such as wind or water) that transfer pollen between plants.

pollutants Chemicals that pollute the environment, such as carbon dioxide, pesticides, fertilizers, industrial chemicals, and discarded medicines.

polychaete A marine annelid worm, such as a clam worm, having paired segmental appendages, trochophore larvae, and separate sexes.

polygenic inheritance The determination of a trait not by just one pair of alleles but by the interaction of many genes.

polymerases Enzymes that enable making copies of DNA.

polymers Large molecules made from chains of similar units, such as the proteins made from amino acids.

polyp In cnidarians, a tall and cylindrical form. Or the body may be a bowl-shaped form called a medusa.

polysaccharides Polymers made of multiple units of sugars.

pongids Chimpanzees and other apes.

population Group of individuals of the same kind living in the same area.

Porifera The animal phylum that contains the sponges. Sponges have pores all over the body, only one cell layer, and only one opening to the body cavity.

positron emission tomography See PET.

potassium An important mineral in plant and animal nutrition. Potassium participates in the propagation of nerve impulses, and one of its isotopes is used in radioactive dating.

Precambrian The era preceding the Paleozoic, from about 3.8 billion to 544 million years ago. Its name comes from Cambrian, the first period in the Paleozoic.

prefrontal cortex Gray matter in the front portion of the cerebrum, where many thought processes take place.

primates Members of the order Primates, which includes lemurs, monkeys, apes, and human beings. Most primates have nails rather than claws; have many teeth; and all have eyes that face forward, allowing stereoscopic vision.

primitive Evolved early; usually used in comparison to a more advanced group.

principle of independent assortment Mendel's discovery that during formation of gametes, the factors for different traits did not stay together but behaved independently.

principle of segregation Mendel's first law, that each individual has pairs of factors for each trait and that the members of the pair separate during formation of gametes.

prions Infective proteins.

profile A series of distinct horizontal soil layers.

progesterone A hormone that prepares the uterus for pregnancy; one of the hormones in birth control pills.

prokaryotes The earliest organisms, in which the cells have no defined nuclei or other membrane-bound organelles in their cells. The prokaryotes and eukaryotes are the two great divisions of all living things.

properties Characteristics.

prophase The first stage of mitosis. The DNA becomes condensed into chromosomes. The nuclear envelope begins to disappear, and a spindle is formed from the centrioles and microtubules.

prophylactically As a preventive.

prosimians Primates such as lemurs, tarsiers, lorises, and bushbabies, which do not resemble humans.

proteinaceous Containing proteins.

proteoses Compounds resulting from the partial hydrolysis of proteins.

Proterozoic The latter part of the Precambrian, from 2.5 billion years ago to 544 million years ago.

protoctists Early eukaryotes; most unicellular, some multicellular; some autotrophic, some heterotrophic.

pseudopods The "false feet" of sarcodines.

Psilophyta Whisk ferns.

psychrophiles Archaeans that can live in very cold climates.

Pterophyta Ferns.

puberty Stage at which a person produces gametes and develops secondary sex characteristics.

punctuated equilibrium The process of long periods of equilibrium being interrupted by relatively short periods of mutations and visible changes in organisms.

purines The class of bases having a distinctive shape that allows them to form two hydrogen bonds. Adenine and guanine are purines.

pyrimidines The class of bases having a distinctive shape that allows them to form three hydrogen bonds. Cytosine and thymine are pyrimidines.

radially symmetrical Arranged around a central axis. When any plane is passed vertically through the axis, the body is divided into two roughly mirror images.

radiating Spreading out.

radioactive dating Fossil-dating method based on the facts that rocks contain radioactive elements and that any radioactive element breaks down to other elements at a constant, predictable rate.

radioactive isotopes Forms of uranium, carbon, and some other elements that are radioactive. Many are used for radioactive dating and as tracers in medical procedures and in biological research.

range Total extent of something; lowest through highest measurements, or the area used by a population of organisms.

reabsorbed Taken up again.

receptor sites Sites on a cell's surface into which a chemical must fit, as a key must fit into a lock, before something happens.

recessive An allele that can determine what a genetic characteristic will be only if it is found on both homologous chromosomes. If a dominant allele is on the other chromosome, that allele determines the characteristic.

reflexes Involuntary actions. They follow stimulation of a receptor that leads to passage of a nerve impulse through a reflex arc and back to an effector.

renal artery Artery entering a kidney.

renal corpuscles Structures in fish kidneys that allow the excretion of water but reabsorb salts.

replicate Copy itself exactly; a DNA strand, for instance.

Reptilia The chordate class that includes dinosaurs, snakes, and alligators. Reptiles usually have four legs; their dry skin is covered with scales; the eggs are amniote; and they are exothermic.

reserves Stored materials, such as oil and gas.

respiration Breathing; or the cellular breakdown of sugars for energy.

respiratory system The system comprising the external nares (nostrils), pharynx, larynx, trachea, bronchi, and lungs, which takes in oxygen and exhales carbon dioxide.

restriction enzyme Enzyme found only in bacteria that cuts a DNA molecule at a certain point. Different restriction enzymes look for different sequences of nucleotides.

restriction fragment length polymorphisms See RFLPs.

retina Layer inside the back of the eyeball on which light is focused by the lens. Information about shapes, amount of light, and color travels to the brain through the optic nerve.

retrovirus RNA virus that can move into the nucleus and be transcribed as DNA, which then becomes part of the cell's DNA, controlling the cell.

reverse transcriptase Enzyme that aids the reverse transcription of RNA into DNA.

RFLPs (restriction fragment length polymorphisms) DNA fragments of different lengths yielded by the action of a restriction enzyme. RFLP differences between human populations have been useful in studies of human evolution and migration.

Rh factor An antigen found in the blood of about 85% of adult humans. An Rh-negative person forms antibodies to antigens in Rh-positive blood; if an Rh-negative woman carries an Rh-positive fetus, she will build up antibodies that can attack a fetus in later pregnancies.

rickets A skeletal disease caused by deficiency of sunlight or vitamin D during childhood and causing bone deformities.

RNA A nucleic acid, usually single-stranded, that transcribes the instructions in nuclear DNA and translates them to assemble proteins in the cytoplasm.

rods In vertebrate animals, receptors that transmit visual information in dim light to the brain through the optic nerve.

root nodules Nodules on legume roots, containing bacteria and cyanobacteria that convert atmospheric nitrogen (N_2) to ammonium ions (NH^{4+}) or nitrate ions (NO^{3-}), which can be taken up by the plants' roots and used for making protein.

rough ER ER that has ribosomes at closely spaced intervals along it.

ruminant A cud-chewing animal that has two or three stomachs, such as cattle or sheep.

sand One of the three chief inorganic ingredients of soil. Sand is made of rock particles that are 0.02–2 mm in diameter and are often quartz.

saprophyte Fungus that breaks down dead organic material to obtain nutrients.

saturated fat Fat in which all the internal carbon atoms have hydrogen side groups. When eaten, it tends to elevate LDLs and increase the danger of plaque formation.

savanna A tropical or semitropical grassland, with scattered or no trees.

scrapie A form of spongiform encephalitis found in sheep.

scrotal sac The sac containing the testes.

secondary sex characteristics Characteristics associated with maleness or femaleness. In males, they include a low voice, enlarged larynx (Adam's apple), and chest hair. In females, they include a relatively high voice, enlarged breasts, and wide pelvis.

secretion Producing and releasing a material, such as saliva or a hormone.

sedimentary rock Rock formed by the accumulation of particles of gravel, sand, and mud.

seed coat Tough outer layer of a seed.

seed-fern Fernlike Mesozoic plants having naked seeds.

seismic Pertaining to vibrations of the Earth.

seminal fluid Semen; sperm cells suspended in a fluid containing glandular secretions.

seminal vesicles Glandular pouches that secrete a fluid containing sugar and protein into the seminal fluid.

seminiferous tubules Tubules in the testes in which sperm cells are produced.

Semitic The language family that includes Hebrew, Aramaic, Arabic, and Amharic; also refers to Jews.

sense organs Organs containing specialized nerve cells that can detect light, sound, and specific chemicals. The eyes, ears, nose, and tongue.

sensory Pertaining to the senses. A sensory neuron carries impulses from receptors toward the brain, spinal cord, or a ganglion.

sensory cortex The part of the cerebral cortex that lies mainly posterior to the central sulcus. Each point on it is linked remotely to some sensory receptor in the body.

sex chromosomes The X and Y chromosomes. Human females have two X chromosomes; human males have one X and one Y chromosome.

sexual reproduction Reproduction that involves two parents and usually results in offspring that are a little different from both parents. The parents produce gametes that unite to form a fertilized egg.

sexual selection A type of selection based on the choice of mates having certain characteristics. When there is sexual selection for a characteristic, the frequency of genes associated with the characteristic will increase in the population.

shrubs Short, woody plants.

silica Silicon dioxide, a mineral found in sand, quartz, and other materials.

silt One of the three chief inorganic ingredients of soil. Silt is made of rock particles that are 0.02 to 0.2 mm in diameter and are often quartz.

skeletal system The system of bones that support and protect soft body tissues.

sleeping sickness A tropical disease caused by *Trypanosoma cruzi,* a parasitic protoctist that lives in human blood.

smooth ER ER that has no ribosomes along it.

sociobiologists Biologists who study behavior and social organization in animals, especially regarding genetics and evolution.

soluble Capable of being dissolved.

speciation Formation of a new species.

species Groups of actually or potentially interbreeding natural populations, which are reproductively isolated from other such groups. Some biologists add the condition that the offspring of any crosses be fertile.

specific epithet The word denoting an organism's species, without its generic name. The same specific epithet may be used for species in different genera.

spermicidal Kills sperm cells.

Sphenophyta Horsetails.

spinal cord The long posterior portion of the nervous system, joined to the brain at the anterior end.

spindle A diamond-shaped structure formed from the centrioles and microtubules during mitosis and meiosis. The spindle fibers attach to the centromeres and lead them to the ends of the cell.

spiracles Pores in the cuticle of mandibulates.

spontaneous generation Life arising from nonlife. The theory of spontaneous generation was disproved in the seventeenth century by Italian biologist Francesco Redi.

sporangia Organs that contain spores.

spore Haploid germ cell that can develop into an adult organism without joining another cell.

staining Dyeing tissues with natural or synthetic stains, making various cell structures stand out when viewed under a microscope.

stamens A flowering plant's male reproductive structures, which ring the carpel. The lower part of a stamen is a filament that ends in an anther.

standard deviation (s) In a graph showing normal distribution, a measure of distance from the mean value. In a normal distribution the probability that any one measurement will fall within one s on either side of the mean is always 68.3%; within two s on either side of the mean, 95%; and within three s on either side of the mean, 99.7%.

statistics A branch of mathematics that deals with collecting, analyzing, and interpreting large quantities of numerical data.

stereochemistry A branch of chemistry that deals with the shapes of molecules and arrangements of atoms within them.

steroids A group of chemicals having a characteristic arrangement of one pentagonal and three hexagonal rings of carbon atoms. Important for their biological activity, they include many hormones and bile acids.

sterols A class of steroids that includes cholesterol and related compounds. Sterols are abundant in the tissues of all organisms except bacteria.

stigma Opening of the carpel in a flower.

stomata Microscopic openings in plant leaves and stems that allow for gas (carbon dioxide) and water uptake and loss.

strata Layers.

struggle for survival A concept proposed by Malthus and others that organisms continually compete for resources. Darwin realized that it was an important component of natural selection and the origin of species.

style The part of a carpel above the ovary in a flower.

substrate Underlying layer, such as the soil under plants; or the substance on which an enzyme acts.

sucrose A disaccharide that is made up of fructose and glucose units.

symbiosis Living together.

synapses The spaces between adjacent neurons, into which neurotransmitters are secreted.

target cells Cells that are affected by the secretions of glands.

taxonomists Biologists who specialize in taxonomy.

taxonomy The classification of organisms.

telophase The last stage of mitosis, during which the cytoplasm is divided between the two parts of the cell as a new nucleus forms at each end.

termination The last stage of translation from mRNA to protein, when a termination codon on mRNA is reached.

termination codon A codon on mRNA (such as UAA) that has no matching anticodons. No more amino acids can be added to the growing protein chain when a termination codon is reached.

testes Male reproductive organs, in which sperm cells are produced.

tetrapods Animals with four legs. Early amphibians were the first tetrapods.

thecodonts Dinosaur ancestors, one of the two major reptile groups in the Triassic.

theory A broad explanation of scientific phenomena, supported by much observation and experiment and guiding new scientific work.

therapsids Mammal-like reptiles, one of the two major reptile groups in the Triassic.

thermophiles Archaeans that can tolerate very high temperatures.

thymine $C_5H_5N_2O_2$, a pyrimidine found in DNA that always pairs with adenine.

tissue fluid Fluid surrounding cells, containing salts dissolved in water.

tissues A group of cells specialized for a common function.

tomography A technique for getting clear X-rays of deep internal structures by focusing on just one plane within the body.

toxins Substances that act as poisons.

tracheae In mandibulates, thin-walled tubes that branch throughout the body and provide gas exchange to all parts of it.

tracts Areas of land, or body systems, or bundles of neurons.

trait Property. Often used as a synonym for *characteristic*. In Mendelian genetics, characteristic is a kind of property, such as eye color, and trait is the specific form of it, such as red or white.

transfer RNA See tRNA.

transformation Griffith's name for the hereditary change from living, harmless bacteria to a disease-producing virulent type caused by extracts from killed encapsulated streptococci.

translocation The transfer of a deleted part of one chromosome to another chromosome, of which it becomes a part.

transpiration Water loss from plants, through stomata in the stems and leaves.

transposons Genes that move from one chromosome to another and disrupt the normal actions of DNA in the cell.

triglycerides A kind of lipid. Nearly all the lipids we eat are triglycerides.

trilobites Some of the earliest arthropods, found often as fossils.

tRNA (transfer RNA) The form of RNA that translates instructions from DNA to protein. It contains anticodons that link to the codons in mRNA; each anticodon adds an amino acid to the growing protein chain.

trochophore larva A free-swimming larva covered with cilia.

turbines Water-driven machines that turn and generate electricity.

tympanum Eardrum.

understory Plants beneath a forest canopy.

undulipodia Tail-like formations that protoctists use for swimming and form parts of cells in more advanced organisms. They superficially resemble the flagella of bacteria but are quite different from them in their detailed structure.

unicellular Having only one cell per organism.

uracil $C_4H_4N_2O_2$, a pyrimidine found in RNA that always pairs with adenine.

urea A nitrogenous waste product that is excreted in urine.

urethra The duct that passes out of the body from the bladder. It carries urine in both sexes and semen in males.

vacuole Membrane-enclosed sac in a cell containing food or water. Vacuoles in plant cells are very large.

vagina Passage to the outside of the body from the uterus; the birth canal.

variable expressivity Differences among family members in how much of a gene's effect can be seen.

variation Differences in genetic characteristics.

vas deferens Tube leading from the testis to the urethra.

vector A carrier. In genetic engineering, a vector is likely to be a plasmid from a microbe; in disease, it is anything that transmits pathogens from one host to another.

venous Pertaining to the veins.

venules Small veins.

vertebrates Chordates that also have vertebrae.

vesicles Membrane-enclosed sacs, similar to but smaller than vacuoles, that transport materials into, out of, and throughout the cell.

vestibule A chamber of the inner ear that contains the apparatus for maintaining equilibrium.

villi Projections into the small intestine that contain the lacteals.

visual field The area that can be perceived by one eye.

vitamins Organic substances needed in small amounts by the body for various metabolic processes that must be either synthesized by intestinal bacteria or obtained in foods or supplements.

volatile Tending to become a vapor at a low temperature.

water vascular system Unique system in echinoderms that consists of canals that radiate into the arms or other body sections, carrying sea water, and protrude through the body wall as tube feet (structures that are adapted for collecting food, acting as chemoreceptors, secreting mucus, and respiring, as well as for moving).

weathering Breakdown by wind and precipitation.

white matter Neural tissue containing myelinated nerve tracts. Found on the surface of the spinal cord and in the interior of the cerebrum.

X chromosomes Sex chromosomes that are found in pairs in a female's body cells and singly in a male's. A male's cells contain a small Y chromosome instead of an X.

X-linked Carried on the X chromosome. X-linked traits are usually seen only in males; only in the rare event that a female inherits the allele from both her mother and her father will she have the disorder.

xerophthalmia A condition in which the eyes are very dry and lusterless, often resulting from a vitamin A deficiency.

xylem Conducting vessels in vascular plants, through which water travels upward from the roots to the rest of the plant.

Y chromosome Sex chromosomes that are found only in a male's body cells. A female's cells contain two X chromosomes; a male's cells contain a small Y chromosome and one X chromosome.

Zygomycota The group of fungi that includes *Rhizopus;* they resemble algae in size and shape.

zygote A fertilized egg, which develops into an embryo.

Selected Bibliography

Boorstin, D. J. *The Discoverers.* New York: Random House, 1983.

Cavalli-Sforza, L. L. *Genes, People, and Language.* New York: North Point Press, 2000.

Darwin, Charles. *The Origin of Species.* London: John Murray, 1859.

———. *The Voyage of the 'Beagle.'* 1845; London: Heron Books.

Eiseley, Loren. *Darwin's Century.* New York: Doubleday, 1958.

Eldredge, N., and S. J. Gould. "Punctuated Equilibria: An Alternative to Phyletic Gradualism." In *Models in Paleobiology,* ed, T. J. Schopf. San Francisco: Freeman, Cooper, 1972.

Flint, R. F., and B. J. Skinner. *Physical Geology,* 2nd ed. New York: Wiley, 1977.

Gould, S. J. *The Structure of Evolutionary Theory.* Cambridge, Mass.: Belknap Press of Harvard University Press, 2002.

———. *The Panda's Thumb.* New York: Norton, 1992.

———. "The Piltdown Conspiracy." *Natural History,* August 1980, 8.

Huxley, Aldous. *Brave New World.* New York: Harper & Row, 1946.

Jensen, W. A. et al. *Biology.* Belmont, Calif.: Wadsworth, 1979.

Krebs, R. E. *Scientific Laws, Principles, and Theories.* Westport, Conn.: Greenwood Press, 2001.

Lehninger, A. L. *Biochemistry.* New York: Worth, 1972.

Maddox, B. *Rosalind Franklin: The Dark Lady of DNA.* New York: HarperCollins, 2002.

Margulis, Lynn, and K. V. Schwartz. *Five Kingdoms: An Illustrated Guide to the Phyla of Life on Earth,* 3rd ed. New York: Freeman, 1998.

Mayr, E. *This Is Biology.* Cambridge, Mass: Belknap Press of Harvard University Press, 1997.

Mokdad, I. H. et al. "The Spread of the Obesity Epidemic in the United States, 1991–1998." *Journal of the American Medical Association* 282, no. 16 (October 27, 1999).

Murray, D. L., and J. A. Bond. *An Experience with Populations.* Reading, Mass.: Addison-Wesley, 1971.

Nash, Roderick. *Wilderness and the American Mind.* New Haven: Yale University Press, 1973.

Penfield, Wilder. *The Mystery of the Mind.* Princeton, N.J.: Princeton University Press, 1975.

Plague and Public Health in Renaissance Europe Project. www.iath.virginia.edu/osheim/plaguein.html.

Raven, P. H., and G. B. Johnson. *Biology,* 2nd ed. St. Louis: Times Mirror College Publishing, 1989.

Science News 159, no. 62 (January 27, 2001).

Steinbeck, J., and E. Ricketts. *Sea of Cortez.* Mount Vernon, N.Y.: Paul P. Appel, 1941.

Taber, W. A., and R. A. Taber. *The Impact of Fungi on Man.* Chicago: Rand McNally, 1967.

The American Museum of Natural History. www.amnh.org/nationalcenter/expeditions/blacksmokers/life_forms.html.

Thomas, Lewis. *The Lives of a Cell.* New York: Viking Press, 1974.

U.S. Department of Agriculture. *Soil: The 1957 Yearbook of Agriculture.* Washington, D.C.: U.S. Government Printing Office, 1957.

Vandervoort, F. S. "A Green Centennial." *American Biology Teacher* 61 (1999): 648.

Watson, J. D. *The Double Helix.* New York: Atheneum, 1968.

White, T., et al. "Pleistocene *Homo sapiens* from Middle Awash, Ethiopia." *Nature* 423 (2003): 742.

Wilson, E. O. *Sociobiology.* Cambridge, Mass.: Belknap Press of Harvard University Press, 1975.

Winchester, Simon. *The Map that Changed the World.* New York: HarperCollins, 2001.

For Further Reading

Many periodicals and newspapers are good sources of information about biology. These are particularly authoritative and readable:

New York Times (especially the Science Times section, published on Tuesdays, which is devoted to the latest findings in the sciences)
Nature
Scientific American
American Scientist
Science News
Audubon
Natural History
Smithsonian (some articles on American science; also contains articles on American art and history)

The articles and books listed next provide additional background about many of the topics in this book:

Aldridge, Susan. *The Thread of Life.* Cambridge: Cambridge University Press, 1996. Modern genetics.

Alper, J. Rethinking Neanderthals. *Smithsonian* 34 (2003): 82. A new look at the Neanderthals.

Angier, N. "Genome Shows Evolution Has an Eye for Hyperbole." *New York Times,* February 13, 2001, p. D1. Redundancy in DNA.

Asimov, Isaac. *A Short History of Biology.* New York: Doubleday, 1964. Early biological history.

Balter, Michael. "Speech Gene Tied to Modern Humans." *Science* 297 (2002): 1105. The origins of speech.

Bates, Henry Walter. *The Naturalist on the River Amazons.* 1864; Berkeley: University of California Press, 1962. Butterflies and other tropical species.

Bates, Marston. *The Forest and the Sea.* New York: Vintage Books, 1960. Ecology and the economy of nature.

Beattie, Andrew, and Paul R. Ehrlich. *Wild Solutions: How Biodiversity Is Money in the Bank.* New Haven: Yale University Press, 2001. Biodiversity as Earth's capital.

Bibliothèque Nationale du Québec. *La Faune du Québec et son Habitat.* Québec: Gouvernement du Québec, 1988. Animals and ecology of Quebec.

Black, J. G. *Microbiology.* Upper Saddle River, N.J.: Prentice Hall, 1996. An introduction to microbiology.

Blunt, Wilfrid. *Linnaeus: The Compleat Naturalist.* Princeton, N.J.: Princeton University Press, 2002. A biography of the father of taxonomy.

Boorstin, D. J. *The Discoverers.* New York: Random House, 1983. Science history.

"Bringing Ordinary Doctors into the Genetics Party." *British Medical Journal* 322 (2001): 1016. Genetics and the future of medicine.

Brody, J. E. "Fibromyalgia: Real Illness, Real Answers." *New York Times,* August 1, 2000. A widespread, underrated illness.

———. "Health Sleuths Assess Homocysteine as Culprit." *New York Times,* June 13, 2000. Preventing arterial disease.

Bronowski, Jacob. *The Ascent of Man.* Boston: Little, Brown, 1973. A history of scientific thought.

Buchsbaum, Ralph. *Animals without Backbones.* Chicago: University of Chicago Press, 1976. Invertebrates.

Buffetaut, E., and J.-C. Rage. "Fossil Amphibians and Reptiles and the Africa–South America Connection." In *The Africa–South America Connection,* ed. W. George and R. Lavocat. Oxford: Clarendon Press, 1993. Fossil evidence for plate tectonics.

Carson, Rachel. *Silent Spring.* Boston: Houghton Mifflin, 1962. The book that made the public aware of the dangers of pollution in the 1960s.

Cavalli-Sforza, L. L. *Genes, People, and Language.* New York: North Point Press, 2000. The connections between genes, languages, and populations, described by a leading geneticist.

Courtillot, V. *Evolutionary Catastrophes: The Science of Mass Extinction.* Cambridge: Cambridge University Press, 1999. Extinctions.

Crick, F. "The Double Helix: A Personal View." *Nature* 248 (1974): 766–69. Crick's recollections of his work with Watson on DNA.

Dart, Raymond A. *Adventures with the Missing Link.* Philadelphia: Institutes Press, 1967. A firsthand account of the first australopithecine fossil discovery.

Darwin, Charles. *The Origin of Species.* London: John Murray, 1859. One of the most important biology books ever written.

———. *The Voyage of the 'Beagle.'* New York: Harper, 1959. Darwin's own account of the great voyage of his life.

———. *The Illustrated Origin of Species.* Abridged by Richard Leakey. New York: Hill and Wang, 1979. A very readable version of Darwin's major book, abridged by a major anthropologist.

Dawkins, Richard. *The Selfish Gene.* New York: Oxford University Press, 1976. The first book to popularize the idea of kin selection.

Daws, Gavan, et al. *The Illustrated Atlas of Hawaii.* Honolulu: Island Heritage, 1970. Hawaiian history, geography, and biology.

de Kruif, Paul. *Microbe Hunters.* New York: Harcourt Brace, 1926. Early bacteriologists.

Dethier, Vincent. *To Know a Fly.* San Francisco: Holden Day, 1962. Animal behavior.

Edwards, L. E., and J. Pojeta Jr. *Fossils, Rocks, and Time.* Washington, D.C.: U.S. Geological Survey, 1994. Geology, radioactive dating.

Eicher, D. L. *Geologic Time.* Englewood Cliffs, N.J.: Prentice Hall, 1968. Geology and evolution.

Eiseley, Loren. *Darwin's Century.* New York: Doubleday, 1958. Darwin in historical context; written by a leading science writer and anthropologist.

———. *The Immense Journey.* New York: Random House, 1957. An engrossing book about evolution.

Eldredge, N., and S. J. Gould. "Punctuated Equilibria: An Alternative to Phyletic Gradualism." In *Models in Paleobiology,* ed. T. J. Schopf. San Francisco: Freeman, Cooper, 1972, pp. 82–115. The original article about punctuated equilibrium.

Eldredge, Niles. *The Pattern of Evolution.* New York: Freeman, 1999. Punctuated equilibrium.

Garrett, L. *The Coming Plague.* New York: Farrar, Straus and Giroux, 1994. The dangers of antibiotic-resistant bacteria.

Gould, S. J. *The Structure of Evolutionary Theory.* Cambridge, Mass.: Belknap Press of Harvard University Press, 2002. Gould's last book.

———. *The Panda's Thumb.* New York: Norton, 1992. How evolution happens, by a prominent evolutionary biologist.

———. "The Piltdown Conspiracy." *Natural History,* August 1980, p. 8. Discovery of a fraud.

Greene, Harry. *Snakes.* Berkeley: University of California Press, 1997. A reference book by a leading herpetologist.

Gross, Michael. *Travels to the Nanoworld.* Boulder, Colo.: Perseus Publishing, 2001. The nanotechnology revolution.

Hardin, G. "Living in a Lifeboat." *Bioscience* 24 (1974): 561. Hardin's controversial lifeboat ethic.

Howland, John L. *The Surprising Archaea.* New York: Oxford University Press, 2000. The archaeans and their place in the biosphere.

Huxley, Aldous. *Brave New World.* New York: Harper & Row, 1946. Science fiction predictions that have proved to be partly accurate.

Johanson, R., and M. Edey. *Lucy: The Beginnings of Humankind.* New York: Simon and Schuster, 1981. Lucy and how she fits into the picture of human evolution.

Kahn, J. "The Science and Politics of Super Rice." *New York Times,* October 22, 2002. Politics and genetically altered organisms.

Karp, Walter. *Charles Darwin and the Origin of Species.* New York: American Heritage, 1968. A biography of Darwin.

Keller, Evelyn Fox. *A Feeling for the Organism.* San Francisco: Freeman, 1983. A biography of Barbara McClintock.

———. *The Century of the Gene.* Cambridge, Mass.: Harvard University Press, 2000. How knowledge about genetics has advanced and how it has influenced our perception of biology.

Kolata, G. "Using Genetic Tests, Ashkenazi Jews Vanquish a Disease." *New York Times,* February 18, 2003. Controlling Tay-Sachs.

Krebs, R. E. *Scientific Development and Misconceptions through the Ages.* Westport, Conn.: Greenwood Press, 1999. Some surprising information about the history of science.

———. *Scientific Laws, Principles, and Theories.* Westport, Conn.: Greenwood Press, 2001. An overview of the laws underlying scientific phenomena and the theories explaining them.

Ladd, Gary. *Geology of the Grand Canyon.* Grand Canyon, Ariz.: Grand Canyon Natural History Association, 1984. The record of evolution, in the rocks.

Leopold, Aldo. *A Sand County Almanac.* London: Oxford University Press, 1949. An ecology classic.

Levinton, J. S. "The Big Bang of Animal Evolution." *Scientific American* 267 (1992): 84–93. The Cambrian explosion.

Maddox, B. *Rosalind Franklin: The Dark Lady of DNA.* New York: HarperCollins, 2002. The biography of Franklin, who failed to get the credit she deserved for contributing to the Watson-Crick model of DNA.

Margulis, Lynn, and Karlene Schwartz. *Five Kingdoms: An Illustrated Guide to the Phyla of Life.* New York: Freeman, 1998. Descriptions of all the phyla, with emphasis on the five-kingdom model.

Maynard Smith, J. *Evolution and the Theory of Games.* Cambridge: Cambridge University Press, 1982. The biological use of game theory.

Mayr, Ernst. *This Is Biology.* Cambridge, Mass.: Belknap Press of Harvard University Press, 1997. Evolution by an expert in speciation.

McMichael, Tony. "The Biosphere, Health, and 'sustainabaility.'" *Science* 297 (2002): 1105. Maintaining both human health and planetary health.

Milius, S. "Tougher Weeds? Borrowed Gene Helps Wild Sunflower." *Science News* 162 (2002): 99. A possible adverse effect of genetic engineering.

Milne, Lorus, and Margery Milne. *National Audubon Society Field Guide to North American Insects and Spiders.* New York: Random House, 2000. An authoritative field guide.

Morris, Desmond. *The Mammals.* New York: Harper & Row, 1965. A survey of the mammals.

Muir, John. *My First Summer in the Sierra.* Boston: Houghton-Mifflin, 1911. Muir's own account of his introduction to the Sierra Nevada Mountains.

Nash, Roderick. *Wilderness and the American Mind.* New Haven: Yale University Press, 1973. Changing perceptions of wilderness in the United States.

National Academy of Sciences. *Science and Creationism: A View from the National Academy of Sciences,* 2nd ed. Washington, D.C.: National Academy Press, 1999. Evidence supporting biological evolution.

Netting, Jessa. "Finding the Mother of All Hormones." *Science News* 160 (2001): 94. Water pollution and hormone receptors.

Pennisi, E. "Birds Weigh Risk before Protecting Their Young." *Science* 292 (2001): 414–15. An example of game theory.

Peterson, Roger Tory. *Birds over America.* New York: Dodd, Mead, 1964. An introduction to ornithology.

Petrides, George. *A Field Guide to Trees and Shrubs.* Boston: Houghton-Mifflin, 1986. Trees and shrubs of Northeast and north-central United States and southeast and south-central Canada.

Pollack, A. "The Birth, Death and Rebirth of a Novel Disease-Fighting Tool." *Washington Post,* October 3, 2000. Monoclonal antibody research.

Popper, Karl. *The Logic of Scientific Discovery.* New York: Basic Books, 1959. Popper's philosophy.

Raup, D. M. *The Nemesis Affair.* New York: Norton, 1986. One theory about the dinosaurs' extinction.

Raven, P. H., and G. B. Johnson. *Biology.* St. Louis, Mo.: Times Mirror/Mosby, 1989. An excellent college-level biology textbook.

Ricketts, Edward. *Between Pacific Tides.* Stanford, Calif.: Stanford University Press, 1968. Organisms of the intertidal zones.

Scheffer, Victor. *The Year of the Whale.* New York: Scribner's, 1969. The life of a young sperm whale.

Seelye, Katharine Q. "Bush Backs Federal Funding for Some Stem Cell Research." *New York Times,* August 10, 2001. Politics and stem cell research.

Sereno, P. C. "The Evolution of Dinosaurs." *Science* 284 (1999): 2137. How dinosaurs evolved.

Shelford, V. E. *The Ecology of North America.* Urbana: University of Illinois Press, 1963. North American environments.

Smith, Homer W. *From Fish to Philosopher.* New York: Doubleday, 1961. Evolution with emphasis on the kidney and its maintenance of the internal environment.

Steinbeck, John, and Edward Ricketts. *Sea of Cortez.* Mount Vernon, N.Y.: Paul P. Appel, 1941. An entertaining account of the Sea of Cortez expedition.

Stewart, Amy. *From the Ground Up—The Story of a First Garden.* New York: Algonquin Books, 2001. A book for a beginning gardener.

Stone, Irving. *The Origin.* New York: Doubleday, 1980. An absorbing biography of Charles Darwin.

Strick, James E. *Sparks of Life: Darwinism and the Victorian Debates over Spontaneous Creation.* Cambridge, Mass.: Harvard University Press, 2000. A history of the spontaneous generation controversy and research on the origin of life.

Stringer, C. "Human Evolution: Out of Ethiopia." *Nature* 423 (2003): 692. White's discovery.

Taber, W. A., and R. A. Taber. *The Impact of Fungi on Man.* Chicago: Rand McNally, 1967. An overview of fungi.

Thomas, Lewis. *The Lives of a Cell.* New York: Viking Press, 1974. Cellular biology, by an engaging medical writer.

Turnbull, Colin M. *The Forest People.* New York: Simon and Schuster, 1962. The Pygmies of the Congo.

U.S. Department of Agriculture. *Soil: The 1957 Yearbook of Agriculture.* Washington, D.C.: U.S. Government Printing Office, 1957. Part of a classic series published by the USDA.

Wade, N. (ed.). *The Science Times Book of Genetics.* New York: Lyons Press, 1998. Recent research in genetics.

Wade, N. "The human Family Tree: 10 Adams and 18 Eves." *New York Times,* May 2, 2000. Human origins.

———. "Y Chromosome Bears Witness to Story of the Jewish Diaspora." *New York Times,* May 9, 2000. Y chromosome research.

Walbauer, Gilbert. *Millions of Monarchs, Bunches of Beetles.* Cambridge, Mass.: Harvard University Press, 2000. Adaptive group behaviors of insects and some other animals.

Watson, J. D. *A Passion for DNA.* Plainview, N.Y.: Cold Spring Harbor Laboratory Press, 2000. Watson's late-in-life story about his work with DNA.

———. *The Double Helix.* New York: Atheneum, 1968. Watson's hilarious behind-the-scenes account of how the Watson and Crick DNA model was achieved.

Weiner, Jonathan. *The Beak of the Finch.* New York: Knopf, 1994. Ongoing evolution.

Whitaker, John. *The Audubon Society Field Guide to North American Mammals.* New York: Knopf, 1980. Mammals of North America.

White, T., et al. "Pleistocene *Homo sapiens* from Middle Awash, Ethiopia." *Nature* 423 (2003): 742. The oldest *Homo sapiens* so far.

Wilford, John N. "Tests Suggest Neanderthals Were Hunters, Not Scavengers." *New York Times,* June 13, 2000. How Neanderthals lived.

———. "'Y' Fossils May Be Earliest Human Link." *New York Times,* July 12, 2001. Recent research on early humans.

Wills, Christopher, and Jeffrey Bada. *The Spark of Life: Darwin and the Primeval Soup.* Boulder, Colo.: Perseus Publishing, 2000. Hypotheses about the origin of life and its possible future.

Wilson, Don E., and Sue Ruff (eds.). *The Smithsonian Book of North American Mammals.* Washington, D.C.: Smithsonian Institution Press, 1999. Data on more than 400 North American mammals.

Wilson, E. O. *Sociobiology.* Cambridge, Mass.: Belknap Press of Harvard University Press, 1975. A major book about sociobiology.

Winchester, Simon. *The Map that Changed the World.* New York: HarperCollins, 2001. The story of Smith's discoveries and of his shabby treatment by the scientific establishment of the day.

Woese, Carl R. "On the Evolution of Cells." *Proceedings of the National Academy of Sciences* 99 (2002): 8742–47. Woese's three-kingdom system of classification.

Wong, Kate. "Who Were the Neanderthals?" *Scientific American* 282 (2000): 99–107. A new look at the Neanderthals.

Yoon, C. K. "Sniffing Out the Relatives, for Survival's Sake." *New York Times,* May 2, 2000. Sensory adaptation.

———. "When Biological Control Gets out of Control." *New York Times,* March 6, 2001. The limits of biological controls.

Zimmer, Carl. "All for One and One for All." *Natural History,* February 2, 2002, p. 34. Multicellularity.

———. *Evolution, The Triumph of an Idea.* New York: HarperCollins, 2001. The companion book to the *Evolution* public television series.

Index

Page locators followed by letters indicate a figure (f) or table (t); double letters show more than one of these.

About the Author

CAROL LETH STONE, Ph.D., is a biology writer and editor who has contributed to numerous textbooks and other materials for science education.